S0-CAS-654

BEST SYNTHETIC METHODS

Series Editors

A. R. Katritzky
University of Florida
Gainesville, Florida
USA

O. Meth-Cohn
Sunderland Polytechnic
Sunderland
UK

C.W. Rees
Imperial College of Science and Technology
London, UK

Richard F. Heck, *Palladium Reagents in Organic Syntheses*, 1985
Alan H. Haines, *Methods for the Oxidation of Organic Compounds: Alkanes, Alkenes, Alkynes and Arenes*, 1985
Paul N. Rylander, *Hydrogenation Methods*, 1985
Ernest W. Colvin, *Silicon Reagents in Organic Synthesis*, 1988
Andrew Pelter, Keith Smith and Herbert C. Brown, *Borane Reagents*, 1988
Basil Wakefield, *Organolithium Methods*, 1988
Alan H. Haines, *Methods for the Oxidation of Organic Compounds: Alcohols, Alcohol Derivatives, Alkyl Halides, Nitroalkanes, Alkyl Azides, Carbonyl Compounds, Hydroxyarenes and Aminoarenes*, 1988
H. G. Davies, R. H. Green, D. R. Kelly and S. M. Roberts, *Biotransformations in Preparative Organic Chemistry: The Use of Isolated Enzymes and Whole Cell Systems*, 1989
I. Ninomiya and T. Naito, *Photochemical Synthesis*, 1989
T. Shono, *Electroorganic Synthesis*, 1991
David Crich and William B. Motherwell, *Free-radical Chain Reactions in Organic Synthesis*, 1991
N. Petragnani, *Tellurium in Organic Synthesis*, 1994
T. Imamoto, *Lanthanides in Organic Synthesis*, 1994
A. J. Pearson, *Iron Compounds in Organic Synthesis*, 1994

Iron Compounds in Organic Synthesis

A. J. Pearson

Department of Chemistry
Case Western Reserve University
Cleveland, Ohio 44106-7078, USA

Academic Press

Harcourt Brace & Company, Publishers

London San Diego New York Boston
Sydney Tokyo Toronto

ACADEMIC PRESS LIMITED
24–28 Oval Road
London NW1 7DX

US Edition published by
ACADEMIC PRESS INC.
San Diego, CA 92101

This book is a guide providing general information concerning its subject matter; it is not a procedural manual. Synthesis is a rapidly changing field. The reader should consult current procedural manuals for state-of-the-art instructions and applicable government safety regulations. The publisher and the authors do not accept responsibility for any misuse of this book, including its use as a procedural manual or as a source of specific instructions

A CIP record for this book is available from the British Library

ISBN 0-12-548270-1

This book is printed on acid-free paper

Typeset by Mackreth Media Services, Hemel Hempstead, Herts
and printed in Great Britain by Hartnolls Limited, Bodmin, Cornwall

Contents

Foreword

There is a vast and often bewildering array of synthetic methods and reagents available to organic chemists today. Many chemists have their own favoured methods, old and new, for standard transformations, and these can vary considerably from one laboratory to another. New and unfamiliar methods may well allow a particular synthetic step to be done more readily and in higher yield, but there is always some energy barrier associated with their use for the first time. Furthermore, the very wealth of possibilities creates an information-retrieval problem. How can we choose between all the alternatives, and what are their real advantages and limitations? Where can we find the precise experimental details, so often taken for granted by the experts? There is therefore a constant demand for books on synthetic methods, especially the more practical ones like *Organic Syntheses, Organic Reactions*, and *Reagents for Organic Synthesis*, which are found in most chemistry laboratories. We are convinced that there is a further need, still largely unfulfilled, for a uniform series of books, each dealing concisely with a particular topic from a *practical* point of view—a need, that is, for books full of preparations, practical hints and detailed examples, all critically assessed, and giving just the information needed to smooth our way painlessly into the unfamiliar territory. Such books would obviously be a great help to research students as well as to established organic chemists.

We have been very fortunate with the highly experienced and expert organic chemists, who, agreeing with our objective, have written the first group of volumes in this series, *Best Synthetic Methods*. We shall always be pleased to receive comments from readers and suggestions for future volumes.

A.R.K., O.M.-C., C.W.R.

Preface

The accidental discovery of ferrocene in 1951, by Kealy and Pauson [1], and independently by Miller [2], provided a catalyst for the rapid development of organometallic chemistry in the latter half of the twentieth century. Although quite a large number of metal complexes had been reported in the literature prior to that discovery, the subject was not developed rapidly, probably because of the lack of suitable tools for characterization and structure determination. Almost coincident with the discovery of ferrocene, Orgel, Pauling and Zeiss proposed a bonding model for metal carbonyls, introducing the concept of π-backbonding, which allowed a greater understanding of metal–olefin complexes. The σ-bonded structure originally assigned to ferrocene was later recognized to be incorrect [3], and the correct π-bonded "sandwich" structure for the molecule was proposed.

Since the 1950s the number and nature of organometallic compounds that are known has increased at a bewildering pace. Organoiron compounds have played a central role in the development of the subject. Iron forms stable complexes with a wide range of ligands, making this perhaps the most versatile metal in the Periodic Table. Despite this, the use of organoiron complexes in organic synthesis has been somewhat restricted, compared to, for example, palladium, rhodium, and other metals that are mainly used in catalytic processes. Organic chemists are surprisingly reluctant to delve into the stoichiometric applications of organometallic complexes, possibly because of a misconception that all such compounds are air-sensitive and require special handling techniques. The reader is assured that the vast majority of complexes discussed in this book are very easy to prepare. For example if you look at Chapter 5 structures **8** and **9** you will see complexes that can be made on a scale of 250 g or more. Complex **8** is an oil that is slowly oxidized in air, but can be handled quite easily (e.g., solvents used in extraction are removed on a normal rotary evaporator; the complex can be purified by chromatography on silica gel). Complex **9** has been prepared on large scale in the author's laboratory, as the hexafluorophosphate salt, and has been stored in a normal screw-cap reagent jar for up to two years on the laboratory shelf. The material usually runs out before any decomposition has set in. Indeed, complex **9** is stable to boiling water! But it is very reactive toward nucleophiles, the reaction with $NaCH(CO_2Me)_2$ occurring almost instantaneously at 0°C.

Detailed experimental procedures are given in the text only for those preparations that have been reported in detail as full papers in the primary literature. Procedures that appear only in communications have been written in sufficient detail to allow an experienced practitioner to carry out the reaction, but the author accepts no responsibility for any lack of reproducibility. The aim of the book is to present strategic considerations that may be important in the design of a synthesis, as well as to convince the potential practitioner that most of the experimental chemistry is fairly straightforward. This approach will hopefully encourage the reader to visualize applications of the chemistry at hand to his/her own synthetic problems.

It is also not the purpose of this book to give a detailed discussion of bonding, etc., nor to give an encyclopaedic discussion of organoiron chemistry. The student of fundamental organometallic chemistry will undoubtedly be disappointed with this approach but it is hoped that the synthetic organic chemist will focus on the business end of the molecules (the organic ligands) and begin to appreciate the wide range of applications that are possible.

References

1. T. J. Kealy and P. L. Pauson, *Nature (London)* **168**, 1039 (1951)
2. S. A. Miller, J. A. Tebboth and J. F. Tremaine, *J. Chem. Soc.* 632 (1952)
3. G. Wilkinson, M. Rosenblum, M. C. Whiting and R. B. Woodward, *J. Am. Chem. Soc.* **74**, 2125 (1952)

Detailed Contents

–1–

Iron Carbonyls in Synthesis

There are three stable carbonyls of iron: pentacarbonyliron [$Fe(CO)_5$]; nonacarbonyldiiron [$Fe_2(CO)_9$]; dodecarbonyltriiron [$Fe_3(CO)_{12}$]. In addition, a number of anionic iron carbonyls, such as $Fe(CO)_4^{2-}$, $Fe_2(CO)_8^{2-}$ and $Fe_3(CO)_{11}^{2-}$, their corresponding hydrides, several iron carbonyl halides, and a number of substituted derivatives [such as $Fe(CO)_4PR_3$ and $Fe(CO)_3(PR)_2$] are known. Not all of these have found applications in organic synthesis, and we shall focus only on those complexes that have been used to effect potentially useful reactions. Detailed reviews of the chemistry of iron carbonyls are available[1].

Pentacarbonyliron [$Fe(CO)_5$] was discovered independently by Monde[2] and Berthelot[3] in 1891. It is obtained as a yellow, musty smelling liquid, b.p. *ca* 106°C, by a direct reaction between finely divided iron and carbon monoxide. Since the compound is commercially available and inexpensive, there is little need for its preparation in the laboratory. It is toxic, but its relatively high vapor pressure allows it to be handled (with appropriate precautions) in a standard laboratory fume hood. Removal of unreacted $Fe(CO)_5$ from reaction mixtures is best effected by using a rotary evaporator equipped with a solid CO_2 condenser. Solvents usually distil over into the receiver, while $Fe(CO)_5$ freezes onto the condenser wall. After completion of the operation the $Fe(CO)_5$ can be allowed to thaw and drip into bromine water, whereupon it is destroyed. These procedures should, of course, be conducted in the fume hood.

Diiron nonacarbonyl is prepared as shiny gold platelets by photolysis of a solution of $Fe(CO)_5$ in cold acetic acid, using a standard medium-pressure mercury lamp[4]. In a sunny but no too hot climate, exposure of the stirred $Fe(CO)_5$ solution to sunlight provides an excellent preparative method. Carbon monoxide is evolved during these preparations, so the apparatus should be vented to a fume hood. This material is also commercially available and is received in brown glass reagent jars, with wax-sealed caps. The contents should be inspected very carefully **before** opening the jar, because long storage leads to some deterioration to give $Fe_3(CO)_{12}$ and finely divided metallic iron, evidenced by black patches in the $Fe_2(CO)_9$ sample. The iron impurity is pyrophoric and will ignite if the jar is opened in air (as this author well remembers). If impurities are detected the jar should be opened in an inert atmosphere (N_2 or Ar glove bag) and the iron can be destroyed by careful addition of dilute hydrochloric acid. The iron carbonyl can be recovered by filtration, washing with water, then ethanol and ether, followed by drying *in vacuo*.

Nonacarbonyldiiron is an excellent source of $Fe(CO)_4$ and $Fe(CO)_3$ for the preparation of, for example, diene-$Fe(CO)_3$ complexes under mild conditions (see Chapter 4).

Dodecarbonyltriiron is obtained as a dark-green solid by oxidation of $HFe(CO)_4^-$ using MnO_2[5], or by the oxidation of $Fe_3H(CO)_{11}^{3-}$ or $Fe(CO)_4^{2-}$[6]. Nonoxidative methods for its preparation include the thermal decomposition of nonacarbonyldiiron[7]. This compound also undergoes slow degradation to give pyrophoric iron, and the precautions given for $Fe_2(CO)_9$ should be used when handling old samples of the material.

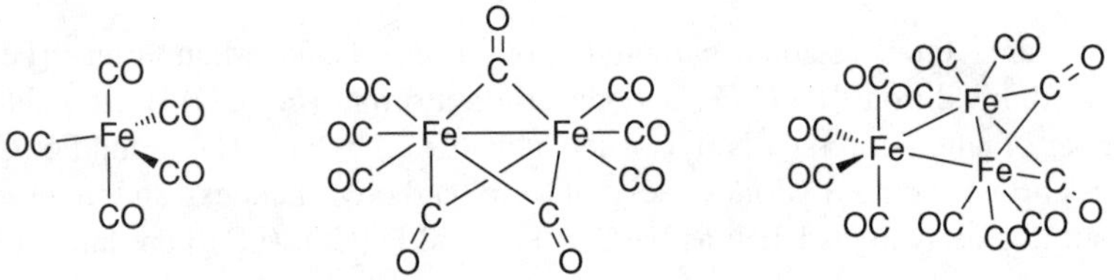

FIG. 1.1. Structures of $Fe(CO)_5$, $Fe_2(CO)_9$ and $Fe_3(CO)_{12}$.

Both $Fe_2(CO)_9$ and $Fe_3(CO)_{12}$ contain a metal–metal bond and bridging carbonyl ligands, and their structures are shown in Fig. 1.1. In addition to their applications in the preparation of alkene-, diene- and dienyl-iron complexes, the metal carbonyls have potential utility as reagents for effecting functional group interconversions. Many of these applications utilize the iron(0) as a reducing agent, and others use the nucleophilic character of iron in, for example, $[HFe(CO)_4]^-$ anion.

1.1 ISOMERIZATION OF OLEFINIC COMPOUNDS

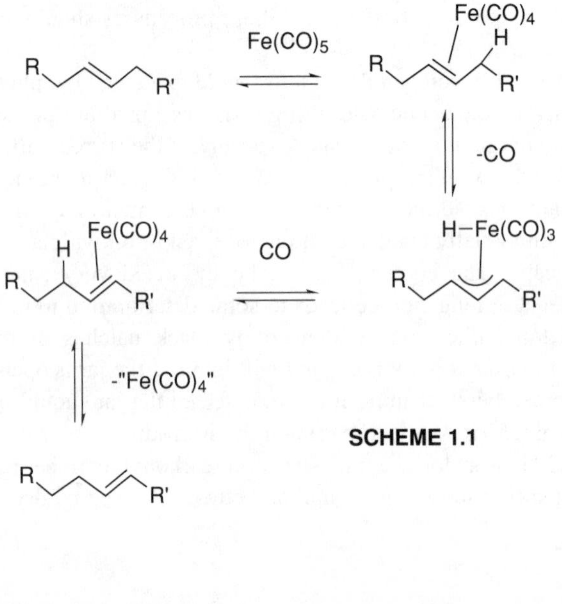

SCHEME 1.1

The reaction of iron carbonyls with alkene derivatives is well known to give alkene-$Fe(CO)_4$ complexes (see Chapter 2). Various kinds of isomerization reaction can occur, typically involving a reversible shift of hydride from the organic ligand to the metal. Often, the presence of another olefinic group allows the eventual formation of a 1,3-diene-$Fe(CO)_3$ complex which represents the most stable arrangement. Removal of the metal then allows the isolation of a product of double bond isomerization. A possible mechanism for C=C bond migration is shown in Scheme 1.1, and a number of examples are given in equations 1.1–1.5[8–12].

Et
$Fe(CO)_5$, Δ
$Fe(CO)_3$
[O]
Et
(1.1)

cat. $Fe(CO)_5$, Δ
(1.2)

CO_2Et
1) $Fe(CO)_5$
2) [O]
CO_2Et
(1.3)

1) $Fe(CO)_5$
2) [O]
R
R
(1.4)

OH
H
H
C_5H_{11}
H OTHP
$Fe_3(CO)_{12}$
DME, 95°C,
30 min
OH
H
H
$Fe(CO)_3$
C_5H_{11}
H OTHP
Steps
O
CO_2H
C_5H_{11}
H OTHP
(1.5)

The example given in equation 1.5 comes from Corey's synthesis of prostaglandin PGC_2[13], and represents one of the few times that $Fe_3(CO)_{12}$ has been used to effect such transformations.

1.2 DEHALOGENATION AND SUBSEQUENT REACTIONS

A number of early investigations showed that reaction of gem-dihalides with $Fe(CO)_5$ results in dehalogenation and coupling to give olefinic compounds[14], two examples being given in equations 1.6 and 1.7. Alper[15] described a reaction of $Fe(CO)_5$ with α-bromoketones which resulted in the formation of coupled 1,4-diketones and debromination products in low yield (equation 1.8).

Ph_2CCl_2 — $Fe(CO)_5$, PhH, Δ; 95% → $Ph_2C{=}CPh_2$ (1.6)

Ar_2CBr_2 — $Fe(CO)_5$, PhH, Δ; 47% → $Ar_2C{=}CAr_2$ Ar = p-$NO_2C_6H_4$ (1.7)

$ArCOC(R^1,R^2)Br$ — 1) $Fe(CO)_5$, DME, Δ; 2) H_2O → $ArCOC(R^1R^2)C(R^1R^2)COAr$ + $ArCOCH(R^1,R^2)$ (1.8)

A breakthrough in terms of synthetic application was made by Noyori[16], who showed that α,α′-dibromoketones and α,α,α′,α′-tetrabromoketones undergo debromination to give reactive oxoallyl cations, presumed to be coordinated to iron, which undergo a very wide range of cycloaddition reactions, examples of which are given in equations 1.9–1.13. Similar reactions had been previously effected by using a zinc–copper couple[17], but these were often low yielding. The use of tetrabromoacetone in the cyclocoupling process overcomes the inability of dibromoacetone itself to undergo reaction with iron carbonyls. The dibromo-derivatives that result from tetrabromoacetone reactions can be debrominated using standard procedures (equations 1.11 and 1.13).

Br ... Br + furan — $Fe_2(CO)_9$, 40 °C, 24 h, 96% → (1.9)

Br ... Br + furan — $Fe_2(CO)_9$ → (1.10)

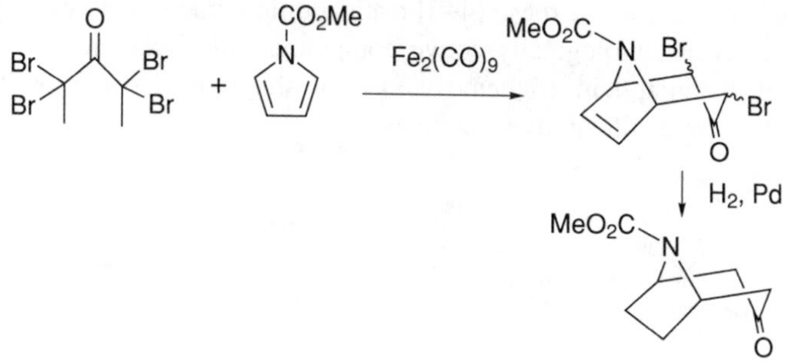

(1.11)

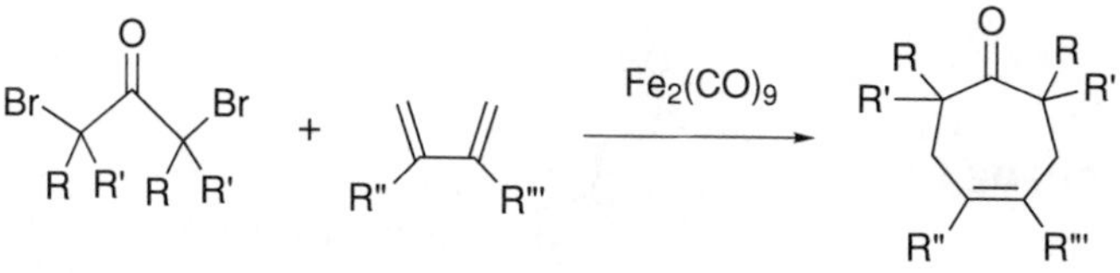

$$R'' = R''' = Me\ (71\%)$$
$$R'' = R''' = H\ (33\%)$$

(1.12)

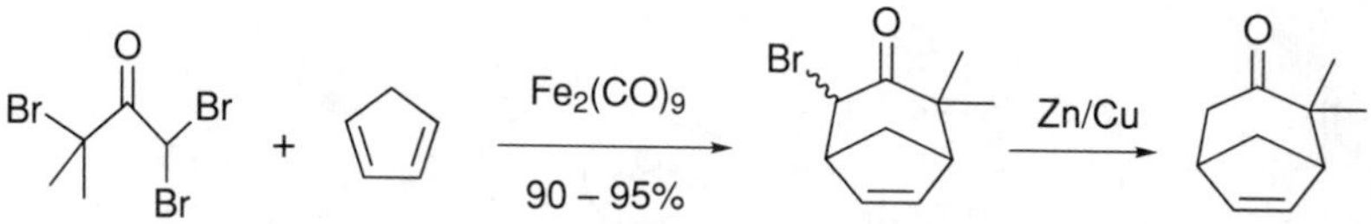

(1.13)

The use of *N*-carbomethoxy pyrrole (equation 1.11) in these reactions allows the construction of tropane alkaloids. It should be noted that the more electron-rich *N*-methylpyrrole does not undergo cycloaddition with the oxoallyl cation, but instead gives the product of electrophilic substitution, a reaction also observed with thiophene (equation 1.14) and with cyclic bromoketones (equations 1.15 and 1.16).

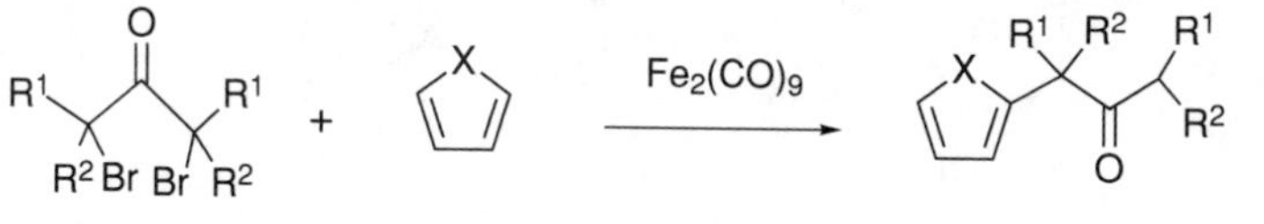

(1.14)

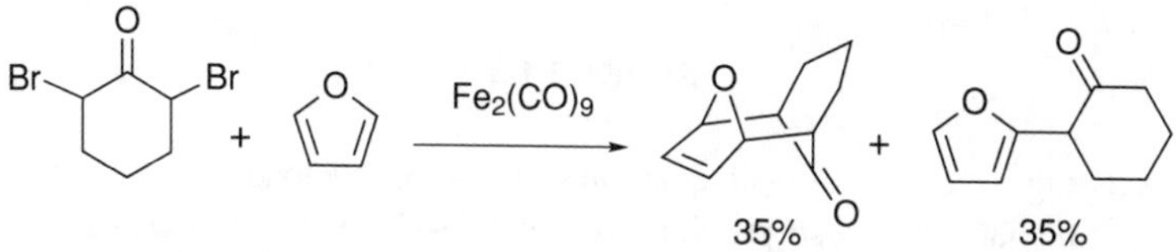

(1.15)

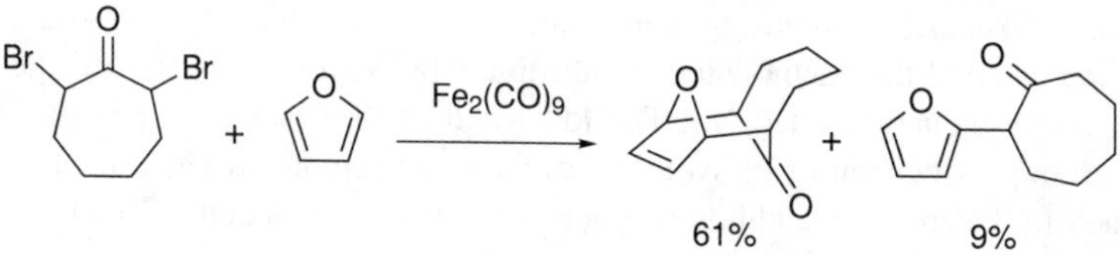

(1.16)

Examples of applications of these [4+3] cycloaddition reactions in the synthesis of natural products and biologically active compounds are found in ref. [16]. Total syntheses of β-thujaplicin (Scheme 1.2), and the *C*-nucleoside pseudouridine (Scheme 1.3) serve as illustrative examples.

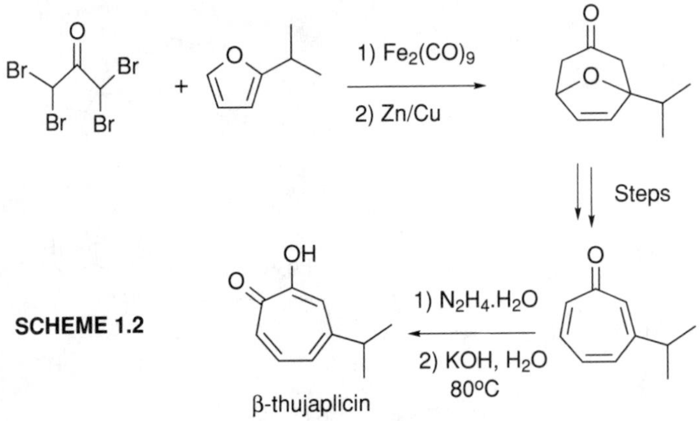

SCHEME 1.2

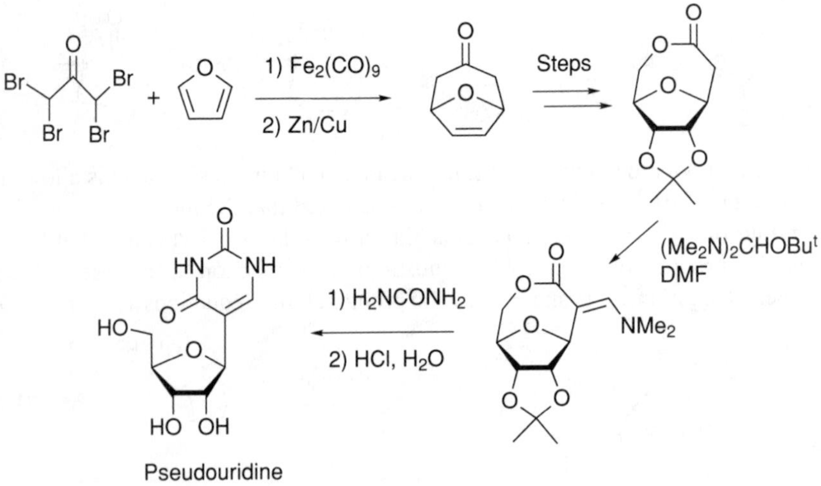

SCHEME 1.3

The preceding reactions are all examples of [4+3] cycloadditions involving an oxoallyl intermediate. It is also possible to effect [3+2] cycloaddition reactions with nucleophilic alkenes, examples of which are shown in equations 1.17–1.21. The synthetic potential of this methodology is illustrated by Noyori's one-step synthesis of (±)–cuparenone (equation 1.19)[18]. The low yield of the reaction is actually better than that from previous multistep syntheses of the same compound[19]. Furanones can be obtained by hetero cycloaddition reaction with amides (equation 1.20). The [3+2] reaction does, however, have limitations, since in some cases the cycloadduct becomes the minor product relative to competing ene reaction products (equation 1.21).

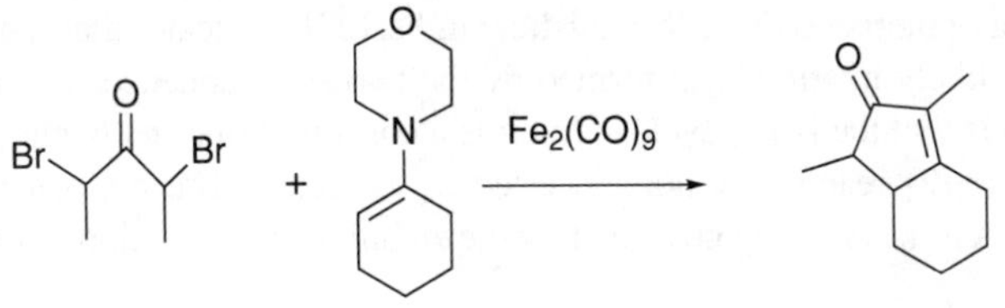

$Fe_2(CO)_9$ (1.17)

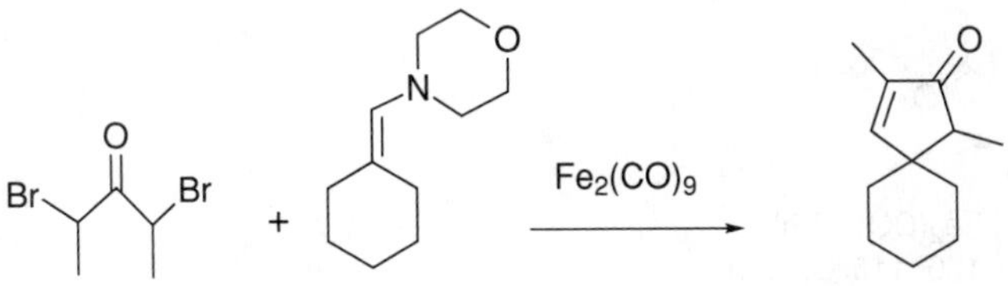

$Fe_2(CO)_9$ (1.18)

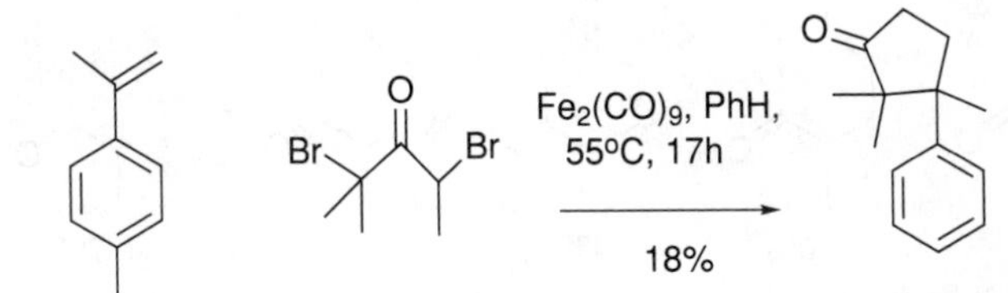

$Fe_2(CO)_9$, PhH, 55°C, 17h, 18% (1.19)

(±)-Cuparenone

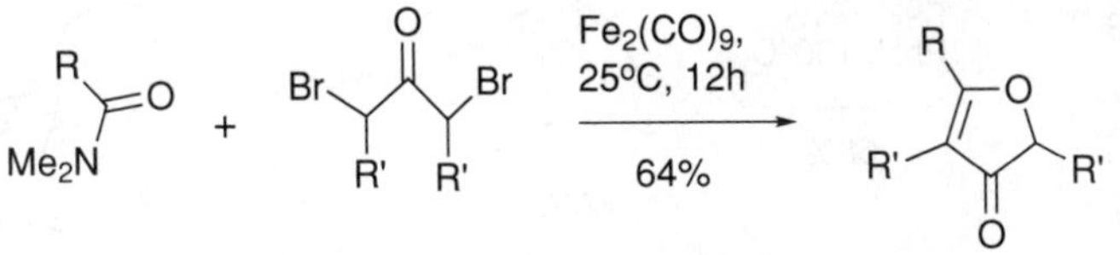

$Fe_2(CO)_9$, 25°C, 12h, 64% (1.20)

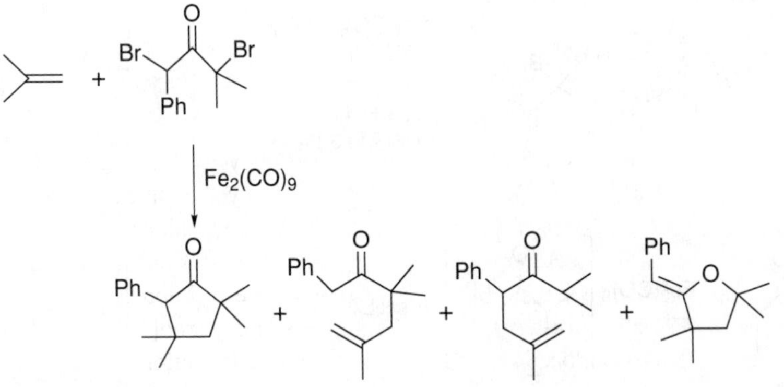

$Fe_2(CO)_9$ (1.21)

Intramolecular coupling of the oxoallyl intermediate with dienes and monoolefins has also been carried out[20,21]. Equations 1.22–1.24 give some examples of this type of reaction, which are interesting because of the relationship of the product structures to various monoterpenoids. Mechanistic studies[22] indicate that an oxoallyl-Fe complex is a likely intermediate, formed by the two step sequence shown in equation 1.25. It appears that the [4+3] cycloaddition is a concerted (thermally allowed) process, whereas the [3+2] reaction is nonconcerted as expected. The regiochemistry of the latter reactions is accordingly governed by the stabilities of the cationic intermediates.

Br Br O
70% $Fe_2(CO)_9$, PhH, 100–110°C, 1.5h
54% + 20% + 4% + 10% + 4–7% (1.22)

Br O Br $Fe_2(CO)_9$, PhH, 100°C 58%
66% + 33% (1.23)

R Br O Br O $Fe_2(CO)_9$, PhH, 80°C, 3h
R = H 41%
R = Me 38%
R O O (1.24)

O Br Br $Fe_2(CO)_9$ [O⁻ Br] [Fe(II)Ln] → [O⁻ +] [Fe(II)Ln] (1.25)

1.3 FUNCTIONAL GROUP INTERCONVERSIONS

Several potentially useful reactions can be effected using iron carbonyls, although in most cases these have little to offer over conventional organic procedures. Some of the processes rely on the reducing capability of the iron carbonyl, such as the deoxygenation reactions in equations 1.26–1.29[23,24]. Desulfurization of episulfides (equation 1.30)[25,26] dehydration of amides and thioamides (equations 1.31 and 1.32)[23,27] and deoximation (equation 1.33)[28] are all potentially useful reactions that have not be fully exploited in synthesis.

$$R_3NO \xrightarrow[45-80\%]{Fe(CO)_5,\ \Delta} R_3N \tag{1.26}$$

$$ArNO_2 \xrightarrow{Fe(CO)_5,\ \Delta} ArN{=}NAr + ArNH_2 \tag{1.27}$$

$$ArNO_2 \xrightarrow{Fe_3(CO)_{12},\ PhH,\ MeOH} ArNH_2 \tag{1.28}$$

$$ArNO \xrightarrow{Fe(CO)_5,\ \Delta} ArN{=}NAr \tag{1.29}$$

$$\text{episulfide } (R^1, R^2, R^3, R^4) \xrightarrow{Fe_2(CO)_9,\ \Delta} R^1R^2C{=}CR^3R^4 \tag{1.30}$$

$$PhC(=X)NH_2 \xrightarrow[X = O\ 32\%;\ X = S\ 64\%]{Fe(CO)_5,\ \Delta} Ph{-}C{\equiv}N \tag{1.31}$$

$$PhC(=X)NHPh \xrightarrow[X = O\ 15\%;\ X = S\ 47\%]{Fe(CO)_5,\ \Delta} Ph{-}CH{=}NPh \tag{1.32}$$

$$\text{HON=(steroidal enone oxime)} \xrightarrow[67\%]{Fe(CO)_5,\ BF_3.Et_2O} \text{O=(steroidal enone)} \tag{1.33}$$

1.4 CARBONYL INSERTION REACTIONS

Transition metal-mediated CO insertion reactions are quite common and form a very important group of synthetic transformations. Iron carbonyls are capable of effecting a number of CO insertion reactions; electrophilic and nucleophilic iron reagents have been developed for these purposes, both of which will be discussed in this section.

Reaction of pentacarbonyliron with organolithium reagents initially occurs by nucleophilic addition to a CO ligand to give an anionic acyliron intermediate. Quenching the reaction with a proton source leads to the formation of aldehydes, while reaction with a suitable alkyl halide yields a ketone (equations 1.34 and 1.35)[29]. Similarly, reaction of $Fe(CO)_5$ with amines leads to *N*-formyl derivatives (equation 1.36)[30]. A plausible mechanism for these types of reaction is given in Scheme 1.4.

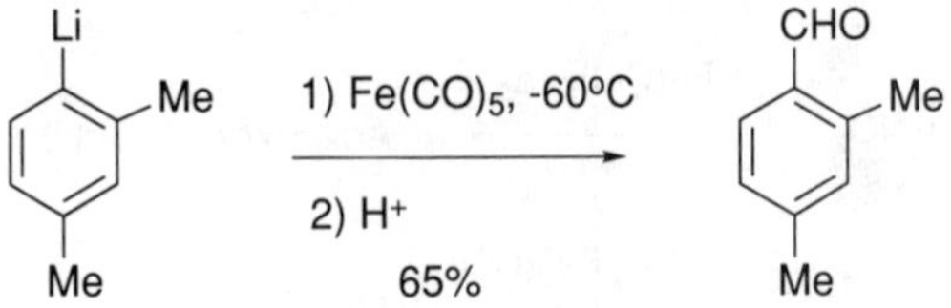

Li, Me, Me 1) $Fe(CO)_5$, -60°C 2) H^+ 65% CHO, Me, Me (1.34)

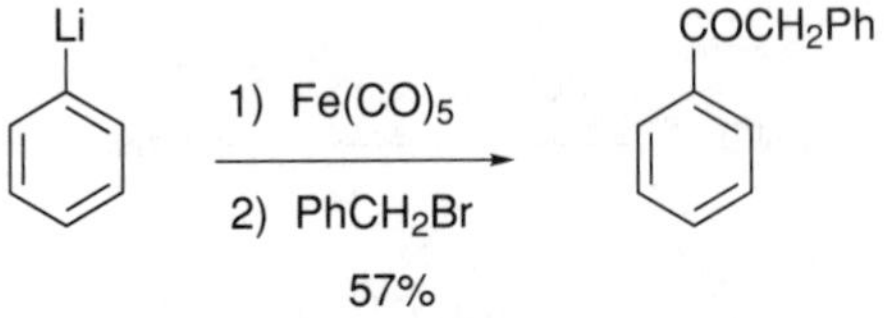

Li 1) $Fe(CO)_5$ 2) $PhCH_2Br$ 57% $COCH_2Ph$ (1.35)

$$R_2NH \xrightarrow{Fe(CO)_5} R_2NCHO \qquad (1.36)$$

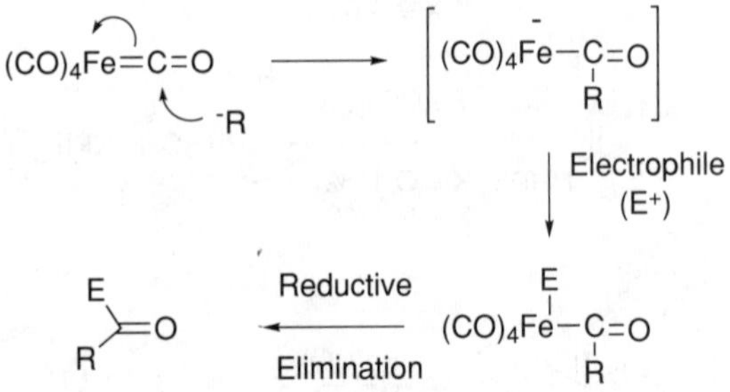

$(CO)_4Fe{=}C{=}O$ ^-R ⟶ $[(CO)_4\overset{-}{Fe}{-}C(R){=}O]$

Electrophile (E^+)

$(CO)_4Fe(E){-}C(R){=}O$ Reductive Elimination ⟶ $E(R)C{=}O$

SCHEME 1.4

Surprisingly, pentacarbonyliron also reacts with electrophilic aryldiazonium salts; when carried out in the presence of aqueous acid, the reaction yields substituted benzoic acids, whereas at lower temperature in dry acetone a diaryl ketone is formed (equation 1.37)[31].

Fe(CO)$_5$, acetone, 5°C ← $N_2^+ Cl^-$ (X-C$_6$H$_4$) → Fe(CO)$_5$, H$^+$, aq. acetone → CO$_2$H (X-C$_6$H$_4$) (1.37)

X = H, Me, OMe, Cl, NO$_2$

Disodium tetracarbonylferrate is an effective formyl or acyl anion equivalent, by virtue of it potent nucleophilicity coupled with the ease with which its alkylation products undergo CO insertion and subsequent reductive elimination. The reagent can be prepared by reduction of $Fe(CO)_5$ with sodium–mercury amalgam or sodium benzophenone ketyl[32]. It is extremely oxygen sensitive and spontaneously ignites in air, but it can be generated and used in solution (e.g., dioxane). Some typical reactions of $Na_2Fe(CO)_4$ are collected in equations 1.38–1.43. The reagent is quite selective; alkyl halides can be converted into aldehydes and ketones in the presence of potentially reactive ester functionality. A change of work-up procedure, using an oxidizing agent and water or an alcohol leads to the formation of carboxylic acids or esters. A plausible mechanism for the CO insertion reaction is outlined in Scheme 1.5.

$$Cl(CH_2)_6Br \xrightarrow[\text{2) AcOH}]{\text{1) } Na_2Fe(CO)_4\text{, CO}} Cl(CH_2)_6CHO \quad 82\% \qquad (1.38)$$

$$n\text{-}C_{12}H_{25}Cl \xrightarrow[\text{2) } O_2\text{, } H_2O]{\text{1) } Na_2Fe(CO)_4\text{, CO}} n\text{-}C_{12}H_{25}CO_2H \qquad (1.39)$$

Br(CH$_2$)$_5$CO$_2$Et → 1) Na$_2$Fe(CO)$_4$, CO; 2) EtI → EtCO(CH$_2$)$_5$CO$_2$Et 74% (1.40)

(steroid ester with Br) → 1) Na$_2$Fe(CO)$_4$, CO; 2) I$_2$, MeOH → (steroid ester with CO$_2$Me) (1.41)

$$Me(CH_2)_4COCl \xrightarrow[\text{2) AcOH}]{\text{1) } Na_2Fe(CO)_4\text{, CO}} Me(CH_2)_4CHO \quad 98\% \qquad (1.42)$$

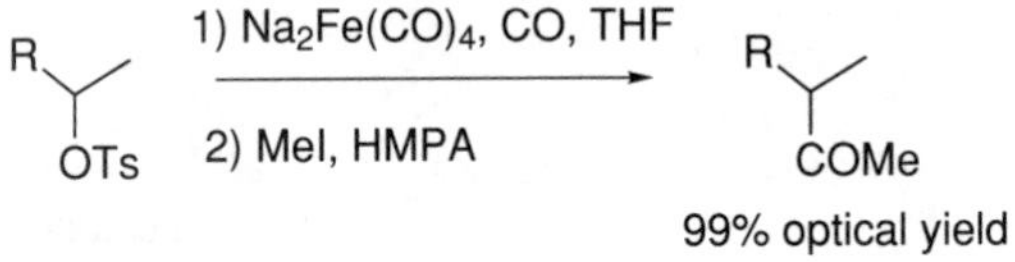

$$\text{R(CH}_3\text{)CH–OTs} \xrightarrow[\text{2) MeI, HMPA}]{\text{1) } Na_2Fe(CO)_4\text{, CO, THF}} \text{R(CH}_3\text{)CH–COMe} \quad \text{99\% optical yield} \qquad (1.43)$$

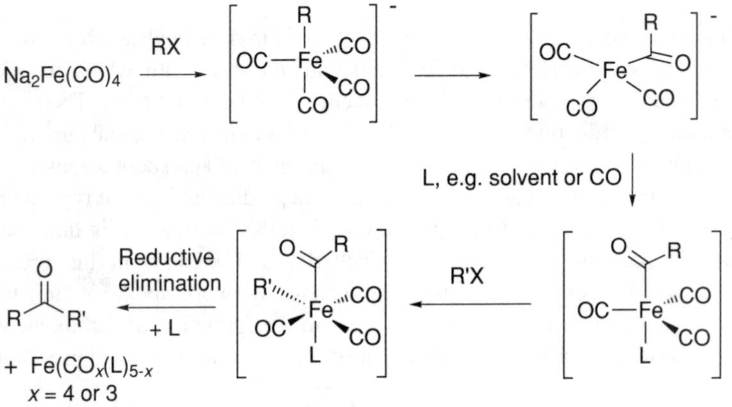

$Na_2Fe(CO)_4$ —RX→ $[RFe(CO)_4]^-$ → $[(RCO)Fe(CO)_3]^-$ —L, e.g. solvent or CO→ $[(RCO)Fe(CO)_3L]^-$ —R'X→ $(RCO)R'Fe(CO)_3L$ —Reductive elimination, + L→ RC(O)R' + $Fe(CO)_x(L)_{5-x}$, x = 4 or 3

SCHEME 1.5

A very interesting and synthetically useful variation on these reactions involves intramolecular trapping of the initial CO insertion product with an alkene, resulting in a carbonylative annulation of ω-bromoalkenes. Five- and six-membered rings are easily formed using this procedure, which is outlined in Scheme 1.6; representative examples are given in equations 1.44–1.46. The reaction shown in equation 1.46 is particularly interesting in that it is a key step in a total synthesis of (±)-aphidicolin[33].

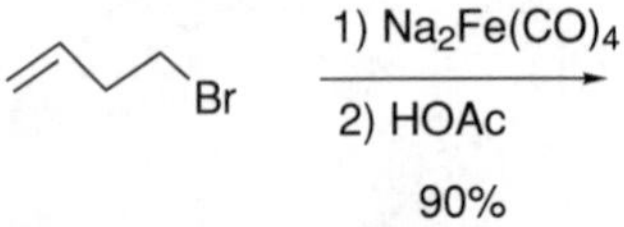

$$CH_2{=}CH(CH_2)_3Br \xrightarrow[\text{2) HOAc}]{\text{1) } Na_2Fe(CO)_4} \text{cyclopentanone} \quad 90\% \qquad (1.44)$$

$$\text{2-methylene-1-(2-tosyloxyethyl)cyclohexane} \xrightarrow[\text{2) HOAc}]{\text{1) } Na_2Fe(CO)_4} \text{2-decalone} \quad 65\%, \; \textit{trans:cis} = 9:1 \qquad (1.45)$$

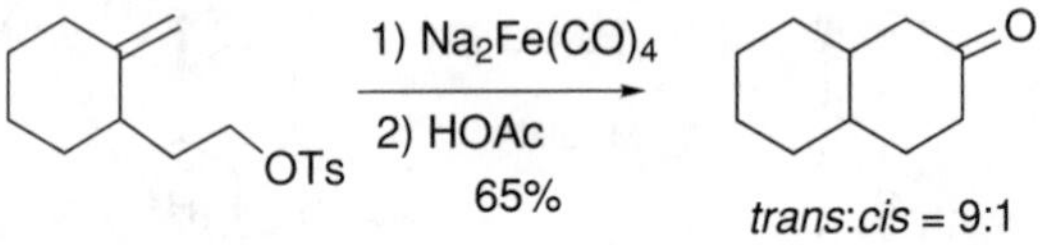

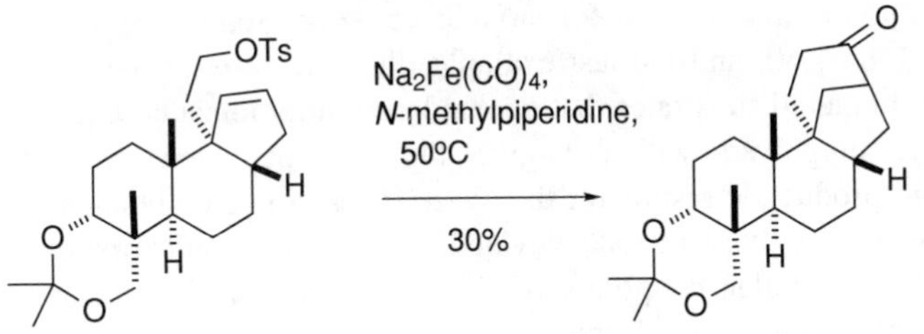

$Na_2Fe(CO)_4$, *N*-methylpiperidine, 50°C; 30% (1.46)

$Na_2Fe(CO)_4$; CO insertion; 16-electron; 18-electron; L; H^+

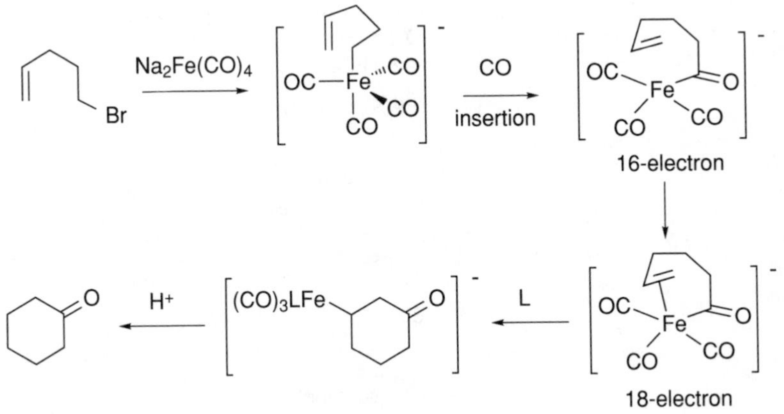

SCHEME 1.6

An intermolecular alkene/alkyl halide carbonylative coupling can be effected when alkenes are used as the olefinic component (equation 1.47), resulting in the initial formation of enone-$Fe(CO)_3$ complexes (see Chapter 4) that are converted into enones by treatment with trimethylamine-*N*-oxide[34].

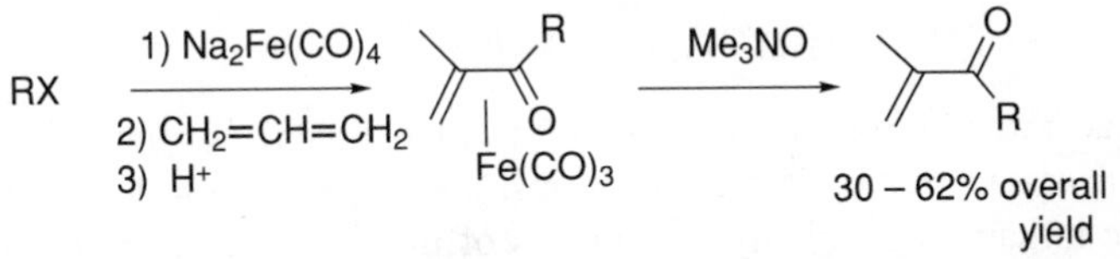

RX → 1) $Na_2Fe(CO)_4$; 2) $CH_2{=}CH{=}CH_2$; 3) H^+ → $Fe(CO)_3$ complex → Me_3NO → 30 – 62% overall yield (1.47)

Recently it has been shown that iron carbonyls can effect an intramolecular [2+2+1] cyclocoupling of alkene, alkyne and CO to give cyclopentenones[35]. This reaction is an iron-promoted equivalent of the now well-known Pauson–Khand reaction, which uses stoichiometric (and more expensive) dicobalt octacarbonyl[36–38]. The intramolecular Pauson–Khand reaction is being increasingly used in organic synthesis, and the iron carbonyl-promoted variant might prove advantageous in terms of cost for large-scale work. Some examples are shown in equations 1.48–1.50, and the last example illustrates one of the limitations of the procedure: hindered substrates that can undergo intramolecular ene reactions at the temperature employed will not give the cyclopentenone carbonyl insertion products. Reproducible results for the [2+2+1] reaction are obtained using $Fe(CO)_5$ in acetonitrile as solvent, at elevated temperature and pressure of carbon monoxide[39]. Similar couplings of diynes to give cyclopentadienone-$Fe(CO)_3$ complexes are discussed in Chapter 4.

Ph
$Fe(CO)_4$(acetone)
toluene, 135°C
CO, 55 psi,
sealed tube
87%
Ph
O
O
(1.48)

EtO_2C
EtO_2C
$Fe(CO)_4$(acetone)
toluene, 135°C
CO, 55 psi,
sealed tube
80%
EtO_2C
EtO_2C
O
(1.49)

EtO_2C
EtO_2C
$Fe(CO)_4$(acetone)
toluene, 135°C
CO, 55 psi,
sealed tube
80%
EtO_2C
EtO_2C
(1.50)

1.5 EXPERIMENTAL PROCEDURES

Preparation of nonacarbonyldiiron

Since nonacarbonyldiiron is commercially available, there is little point in making it in the laboratory for limited applications, but this synthesis does provide some experience in handling carbonyl complexes of iron. Nonacarbonyldiiron is rather expensive, and for large-scale applications, or continual use, it may be preferable to prepare it on a regular basis. The procedure is taken from ref. [4]. A standard Hanovia photochemical reactor may be used in place of the apparatus described herein. Alternatively, the reaction can be carried out in a Pyrex flask set up on the window sill in fairly strong sunlight[40]. *CAUTION: Pentacarbonyliron is very toxic; see p. 90 for methods of handling and disposal. Carbon monoxide is evolved during the reaction, and the procedure should be carried out in an efficient fume hood.* A three-neck 1 liter Pyrex flask equipped with an efficient stirrer, gas inlet, and mercury valve is flushed with nitrogen and charged with pentacarbonyliron (100 ml, 146 g, 0.746 mol) and glacial acetic acid (200 ml). *If the acetic acid contains more than 5% of water, the reaction yields only a brown pyrophoric powder.* The reaction vessel is placed in a 5 liter silvered Dewar flask and is cooled continuously by running water. The stirred reaction mixture is irradiated by a 125 W high pressure mercury lamp. A suitable lamp circuit, constructed to avoid contact of the cooling water with the lamp, is made with a quartz tube (*ca* 30 cm long by 1.7 cm internal diameter). The tube is closed at one end and a Philips mercury vapor lamp (HPK 125 W, type 57203B or its equivalent), to which extension wires with insulated porcelain beads have been attached, is inserted. The lamp assembly is placed in the water bath as close as possible to the reaction vessel. Efficient stirring of the flask contents is essential to avoid deposition of the $Fe_2(CO)_9$ on the sides of the flask, which will reduce light intensity. If deposition does occur, interrupt the reaction occasionally to remove the $Fe_2(CO)_9$ that is formed. The filtrate is then reintroduced into the reaction vessel after it has been flushed with nitrogen once more. If a Hanovia reactor is used the immersion well should be removed and cleaned occasionally for the same reason. After 24 h of irradiation, the $Fe_2(CO)_9$ is removed by filtration and is washed with ethanol and then ether, and dried under vacuum. The yield is usually 100–122 g (74–90% based on $Fe(CO)_5$). The complex is obtained as shiny orange hexagonal leaflets having a density of 2.08 g cm^{-3}. It decomposes at 100–120 °C, is practically insoluble in all organic solvents and is slowly decomposed in THF and methylene chloride. It may be stored for long periods of time in the dark in an inert atmosphere, preferably carbon monoxide. See the comments on aged $Fe_2(CO)_9$ on p. 1.

Isomerization of 1,5-cyclooctadiene (equation 1.2)

The procedure is adapted from ref. [12]. *CAUTION: see comments on pentacarbonyliron in pp. 1 and 90.* Under a nitrogen atmosphere, 1,5-cyclooctadiene (100 g) and pentacarbonyliron (10 g) are stirred and heated at 115 °C for 7 h. The mixture is then cooled to room temperature, filtered and distilled to give unreacted pentacarbonyliron (6 g) and 1,3-cyclooctadiene (94 g, 94% yield).

Iron carbonyl-promoted cycloaddition of α,α′-dibromoketones with 1,3-dienes (equation 1.12)

The procedure is adapted from ref. [41]. To a stirred suspension of nonacarbonyldiiron (4.37 g, 12 mmol) in dry benzene (50 ml), under nitrogen atmosphere, is added, dropwise, a solution of 2,4-dibromo-2,4-dimethylpentan-3-one (2.90 g, 10 mmol) and 2,3-dimethylbutadiene (7.38 g, 90 mmol) in benzene (50 ml). The stirred mixture is heated at 60 °C under a reflux condenser for 40 h. The mixture is cooled and filtered and the solvent, excess diene and $Fe(CO)_5$, that may be formed during the reaction (*CAUTION! see p. 90*), are removed on the rotary evaporator.

The residue is purified by chromatography on silica gel and/or distillation, to give 2,2,4,5,7,7-hexamethyl-4-cycloheptenone (1.38 g, 71% yield).

Iron carbonyl-promoted conversion of oximes to ketones (equation 1.33)[29]

A mixture of the oxime (2–35 mmol) and pentacarbonyliron (1.1 mol per mol of oxime; *CAUTION! see p. 90*) in dry di-n-butyl ether (50–100 ml) containing boron trifluoride etherate (*ca* 5% w/w of oxime) is refluxed with stirring under nitrogen. The progress of the reaction can be followed by TLC. For cyclohexanone oxime the reaction time is 20 h, for cholest-4-en-3-one oxime the time is 16 h. The solution is cooled to room temperature, filtered, and the solvent is removed on the rotary evaporator (for non-volatile ketones only; volatile compounds such as cyclohexanone are isolated as their 2,4-dinitrophenylhydrazones). The residue of crude carbonyl compound is purified by trituration with petroleum ether (b.p. 30–60 °C), hexane or methylene chloride, depending on the solubility properties of the ketone (e.g., for santonin use methylene chloride, for fluorenone use petroleum ether), or by chromatography on Florisil using acetone as eluent (for cholest-4-en-3-one).

Reaction of aryllithium reagents with pentacarbonyliron to give aldehydes (equation 1.34)[29]

Into a 500 ml four-necked flask fitted with a mechanical stirrer, low temperature thermometer, a reflux condenser protected from moisture, a gas bubbler and a dropping funnel, are placed anhydrous ether (50 ml) and, after flushing with dry oxygen-free nitrogen, small pieces of thin lithium foil (1.75 g, 0.25 g atom). The mixture is strirred while a solution of 2,4-dimethylbromobenzene (18.5 g) in anhydrous ether (25 ml) is added dropwise via the dropping funnel over a period of 2.5 h. The internal temperature is maintained at 35 °C throughout the addition. After the addition is complete, stirring is continued for a further 2 h, whereupon the reaction mixture is cooled to –60 °C with a solid CO_2/methanol bath held at –65 °C. Pentacarbonyliron (3.92 g, 0.02 mol) in anhydrous ether (50 ml) is added via the dropping funnel over a period of 2 min, and stirring is continued for a further 2 h while maintaining the internal temperature at –50 °C. The reaction mixture is decomposed by the addition of 95% ethanol (25 ml) followed by 4 N hydrochloric acid (50 ml). The ether layer is separated, washed with saturated aqueous potassium carbonate (3 × 50 ml), then water, dried (Na_2SO_4) and evaporated. The residue is distilled under reduced pressure to give 2,4-dimethylbenzaldehyde (8.7 g, 65% yield), b.p. 110–137 °C at 13 mm Hg. A higher boiling fraction is identified as 2,2′,4,4′-tetramethylbenzyl (0.3 g), b.p. 160–163 °C at 4 mm Hg.

Generation and in situ *reaction of disodium tetracarbonylferrate*

Conversion of 1-bromononane to decanal is an illustrative procedure, taken from ref. [42]. A dry reaction vessel is flushed with dry oxygen-free nitrogen, and is charged with 1% sodium amalgam (2 ml, prepared according to ref. [43], see also p. 32) and dry THF (12 ml). With vigorous magnetic stirring under nitrogen pentacarbonyliron (185 µl, 1.38 mmol) is added, leading to immediate evolution of carbon monoxide, and partial separation of colorless $Na_2Fe(CO)_4$. The mixture is stirred for 1 h at 25 °C and the amalgam is removed through a side arm. A solution of triphenyphosphine (320 mg, 1.2 mmol) in dry THF (2 ml) is added, followed by 1-bromononane (188 µl, 1.0 mmol). The mixture is stirred for 3 h and is then treated with glacial acetic acid (120 µl). After stirring for a further 5 min it is poured into water (100 ml) and extracted with pentane (3×10 ml). The combined pentane extracts are washed with water, dried (Na_2SO_4) and evaporated. The residue is taken up in a small volume of pentane and insoluble iron carbonyl compounds are removed by filtration. The crude product that is obtained after evaporation of the solvent is purified by preparative TLC on alumina (benzene) to give decanal (119 mg, 77% yield), 2,4-DNP m.p. 104–105 °C.

References

1. See, for example: D. F. Shriver and K. Whitmire, in *Comprehensive Organometallic Chemistry*. (ed. G. Wilkinson, F. G. A. Stone and E. W. Abel), Vol. 4, Chapter 31.1.Pergamon Press, Oxford, 1982.
2. L. Monde and F. Quincke, *J. Chem. Soc.* **59**, 604 (1891).
3. M. Berthelot, *C. R. Hebd. Seances Acad. Sci.* **112**, 1343 (1891).
4. E. H. Braye and W. Hubel, *Inorg. Synth.***8**, 178 (1966).
5. R. B. King and F. G. A. Stone, *Inorg. Synth.* **7**, 193 (1963).
6. W. McFarlane and G. Wilkinson, *Z. Anorg. Allg. Chem.* **204**, 165 (1932).
7. H. G. Cutford and P. W. Selwood, *J. Am. Chem. Soc.* **65**, 2414 (1943).
8. H. Alper and J. T. Edward, *J. Organomet. Chem.* **14**, 411 (1968).
9. M. Cais and N. Maoz, *J. Organomet. Chem.* **5**, 370 (1966).
10. G. F. Emerson, Ph. D. Thesis, University of Texas, Austin, Texas, 1964.
11. E. Koerner von Gustorf and J. C. Hogan, *Tetrahedron Lett*, 3191 (1968).
12. J. E. Arnett and R. Pettit, *J. Am. Chem. Soc.* **83**, 2954 (1961).
13. E. J. Corey and G. Moinet, *J. Am. Chem. Soc.* **95**, 7185 (1973).
14. C. E. Coffey, *J. Am. Chem. Soc.* **83**, 1623 (1961).
15. H. Alper and E. C. H. Keung, *J. Org. Chem.* **37**, 2566 (1972).
16. R. Noyori, *Acc. Chem. Res.* **12**, 61 (1979).
17. H. M. R. Hoffmann, K. E. Clemens, E. A. Schmidt and R. H. Smithers, *J. Am. Chem. Soc.* ***94***, 3201 (1972).
18. Y. Hayakawa, F. Shimiza, and R. Noyori, *Tetrahedron Lett.* 1829 (1978).
19. W. Parker, R. Ramage and R. A. Raphael, *J. Chem. Soc.* 1558 (1962).
20. R. Noyori, Y. Hayakawa, M. Funakura, H. Tayaka, S. Murai, R. Kobayashi and S. Tsutsumi, *J. Am. Chem. Soc.* **94**, 7202 (1972).
21. R. Noyori, Y. Hayakawa, H. Tayaka, S. Murai, R. Kobayashi and N. Sonoda, *J. Am. Chem. Soc.* **100**, 1759 (1978).
22. R. Noyori, M. Nishizawa, F. Shimizu, Y. Hayakawa, K. Maruoka, S. Hashimoto, H. Yamamoto and H. Nozaki, *J. Am. Chem. Soc.* **101**, 220 (1979).
23. H. Alper and J. T. Edward, *Can. J. Chem.* **48**, 1543 (1970).
24. J. M. Landesberg, L. Katz and C. Olsen, *J. Org. Chem.* **37**, 930 (1972).
25. R. B. King, *Inorg. Chem.* **2**, 326 (1963).
26. B. M. Trost and S. D. Ziman, *J. Org. Chem.* **38**, 932 (1973).

27. G. Bor, *J. Organomet. Chem.* **11**, 195 (1968).
28. H. Alper and J. T. Edwards, *J. Org. Chem.* **32**, 2938 (1967).
29. M. Rhyang, I. Rhee and S. Tsutsumi, *Bull. Chem. Soc. Jpn.* **37**, 341 (1964).
30. B. J. Bulkin and J. A. Lynch, *Inorg. Chem.* **7**, 2654 (1968).
31. G. N. Schrauzer, *Chem. Ber.* **94**, 1891 (1961).
32. J. P. Collman, *Acc. Chem. Res.* **8**, 342 (1975), and references cited therein.
33. J. E. McMurry, A. Andrews, G. M. Ksander, J. H. Musser and M. A. Johnson, *J. Am. Chem. Soc.* **101**, 1330 (1979).
34. A. Guinot, P. Cadiot and J. L. Roustan, *J. Organomet. Chem.* **128**, C35 (1977).
35. A. J. Pearson and R. A. Dubbert, *J. Chem. Soc., Chem. Commun.* 202 (1991).
36. I. U. Khand, P. L. Pauson, W. E. Watts and M. I. Foreman, *J. Chem. Soc., Perkin Trans.* **1**, 977 (1973).
37. N. E. Schore, *Chem. Rev.* **88**, 1081 (1988).
38. T. R. Hoye and J. A. Suriano, *J. Am. Chem. Soc.* **115**, 1154 (1993). This paper reports a molybdenum carbonyl promoted intramolecular Pauson–Khand reaction.
39. R. A. Dubbert, Ph. D. Dissertation, Case Western Reserve University, 1993.
40. E. Speyer and H. Wolf, *Chem. Ber.* **60**, 1424 (1927).
41. R. Noyori, S. Makino and H. Takaya, *J. Am. Chem. Soc.* **93**, 1272 (1971).
42. M. P. Cooke, Jr, *J. Am. Chem. Soc.* **92**, 6080 (1970).
43. W. B. Renfrow, Jr and C. R. Hauser, *Org. Syn. Coll.* **2**, 607 (1943).

–2–

Alkeneiron Complexes

In general, a transition metal that is π-bound to an olefinic ligand activates the olefin toward nucleophilic attack. This behavior is very common in organometallic chemistry and extends to π-allyl, diene, dienyl and triene or arene complexes. The degree of activation depends to varying extents on the nature of the metal and its oxidation state. It is reasonable to expect that olefin–metal π complexes that are positively charged will be more reactive toward nucleophiles than those which are uncharged. With iron there are examples of both types of alkene complex, and these do indeed follow the expected reactivity patterns.

2.1 NEUTRAL (UNCHARGED) ALKENEIRON COMPLEXES

Replacement of one carbon monoxide ligand of $Fe(CO)_5$ by an alkene in a formal sense, leads to an alkene-$Fe(CO)_4$ complex[1,2]. In actual fact, $Fe_2(CO)_9$ is the reagent of choice for the preparation of these complexes, since it is dissociated to give the reactive, 16-electron $Fe(CO)_4$ species under much milder conditions than is $Fe(CO)_5$. Treatment of ethene under pressure with $Fe_2(CO)_9$ gives ethene-$Fe(CO)_4$ (**2.1**) obtained as a yellow oil, m.p. −21.8°C, b.p. 34°C/12 mm Hg (equation 2.1). The complex can be stored at −80°C for long periods but it decomposes slowly at room temperature to give ethene and $Fe_3(CO)_{12}$. Alkene complexes can also be prepared by photochemical reaction between $Fe(CO)_5$ and the alkene. Most of these complexes undergo rapid oxidation in air to release the olefinic ligand. They show intense infrared absorptions at *ca* 2085, 1990 and 1970 cm^{-1} due to the carbonyl stretching vibrations, and the complexed CC stretch is observed at *ca* 1500 cm^{-1}. This is fairly typical of olefin-$Fe(CO)_x$ complexes, for both monoolefins and dienes.

A variety of substituted alkene-$Fe(CO)_4$ complexes can be prepared. When the substituent is electron withdrawing, as with the examples in equations 2.2 and 2.3, the complexes are more stable and therefore easier to handle than the simple alkene complexes, and many of them are crystalline solids.

$$C_2H_4 + Fe_2(CO)_9 \xrightarrow{\text{50 atm., 2 days}} (H_2C{=}CH_2)Fe(CO)_4 \; (\mathbf{2.1}) + Fe(CO)_5 \qquad (2.1)$$

$Fe_2(CO)_9$, C_6H_6, *ca* 45 °C → Fe(CO)$_4$ (2.2)

2.2 Z = CHO
2.3 Z = CN
2.4 Z = CO_2Me

$Fe_2(CO)_9$, C_6H_6, 45 °C, 4h → Fe(CO)$_4$ (2.3)

2.5

It is usually not possible to isolate alkene-Fe(CO)$_4$ complexes from alkenes that have a leaving group or strained ring in the allylic position, owing to the facility with which such complexes are converted to π-allyl-Fe(CO)$_x$ complexes. In some cases neighboring functionality captures the π-allyl-Fe(CO)$_4$ intermediate to give complexes such as **2.6** and **2.7**. These compounds are interesting for organic synthesis applications and will be discussed further in Chapter 3.

OH, Cl — Fe(CO)$_5$ → Fe(CO)$_3$ (2.4)

2.6

Fe(CO)$_5$, *hν* → Fe(CO)$_3$ (2.5)

2.7

Activation of the alkene toward nucleophilic addition is observed with alkene-Fe(CO)$_4$ complexes, owing to the ability of the iron to act as an electron sink and accommodate the negative charge that builds up. Even ethene-Fe(CO)$_4$ (**2.1**) reacts with fairly gentle nucleophiles such as malonate[3,4]. The intermediate alkyl-Fe(CO)$_4$ anion cannot be isolated, and is usually converted into organic products by appropriate work-up conditions. These are sometimes rather elaborate but generally give good results. When the alkene bears an electron withdrawing group, e.g., complex **2.4**, the product is one of overall Michael addition to the unsaturated ester (equation 2.7). This ability of the Fe(CO)$_4$ group to activate α,β-unsaturated carbonyl compounds has not been fully exploited and so this remains an open area for future investigation.

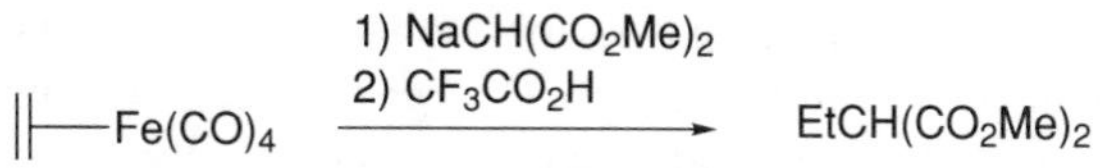

$$\text{2.1: } \| \!-\! Fe(CO)_4 \xrightarrow[\text{2) } CF_3CO_2H]{\text{1) } NaCH(CO_2Me)_2} EtCH(CO_2Me)_2 \quad (45-68\%) \tag{2.6}$$

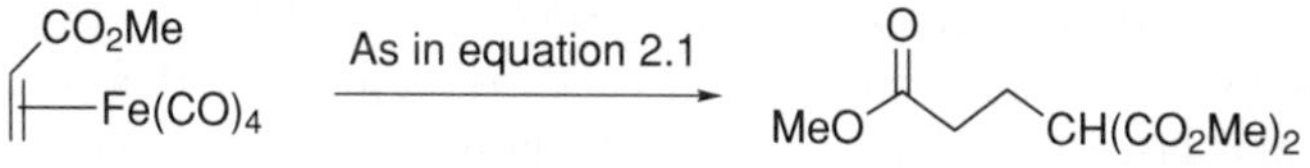

$$\text{2.4: } CO_2Me\text{-alkene}\!-\!Fe(CO)_4 \xrightarrow{\text{As in equation 2.1}} MeO_2C\text{-}CH_2CH_2\text{-}CH(CO_2Me)_2 \quad (91\%) \tag{2.7}$$

If the intermediate alkyl-$Fe(CO)_4$ anion is treated with electrophiles other than proton, e.g., methyl iodide, alkylation on iron occurs, followed by carbonyl migratory insertion and reductive elimination, to give the product **2.8** corresponding to overall Michael addition and acylation of the intermediate enolate (Scheme 2.1).

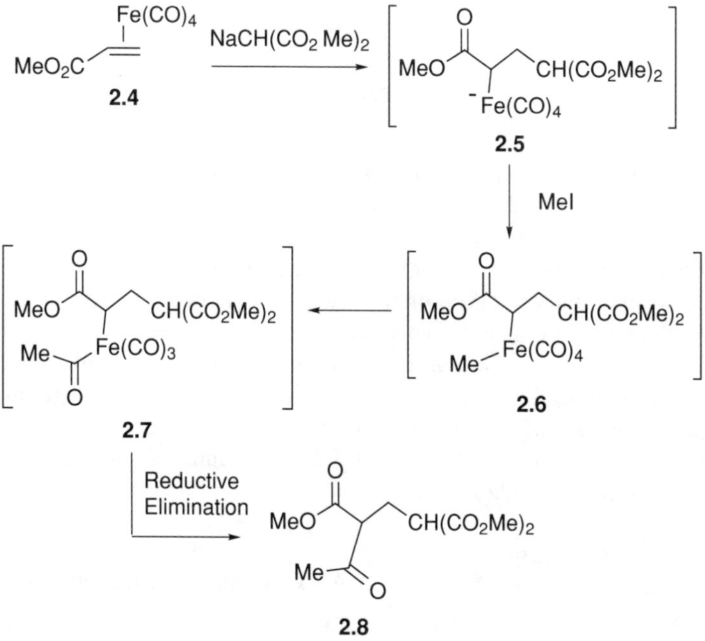

MeO_2C–CH=CH$_2$·$Fe(CO)_4$ (**2.4**) $\xrightarrow{NaCH(CO_2Me)_2}$ [MeO(O=)C–CH($^-Fe(CO)_4$)–CH$_2$–$CH(CO_2Me)_2$] (**2.5**) $\xrightarrow{MeI}$ [MeO(O=)C–CH(Me–$Fe(CO)_4$)–CH$_2$–$CH(CO_2Me)_2$] (**2.6**) ⟶ [MeO(O=)C–CH(MeC(=O)–$Fe(CO)_3$)–CH$_2$–$CH(CO_2Me)_2$] (**2.7**) $\xrightarrow{\text{Reductive Elimination}}$ MeO(O=)C–CH(C(=O)Me)–CH$_2$–$CH(CO_2Me)_2$ (**2.8**)

SCHEME 2.1. Nucleophile addition/acylation sequence for acrylate-$Fe(CO)_4$ complexes.

2.2 CATIONIC ALKENEIRON COMPLEXES

[Alkene-Fe$(CO)_2$Cp]$^+$ complexes (Cp = C_5H_5) have been extensively investigated[5]. The parent ethene complex, as its hexafluorophosphate, is obtained as a pale yellow solid, which is soluble in polar organic solvents such as acetone, acetonitrile and methylene chloride, but insoluble in ether, hexane, etc. This property leads to a convenient method for isolation and purification of the complexes, by precipitation from an appropriate solution using ether. They are appreciably more stable than the alkene-Fe$(CO)_4$ complexes discussed above, and easier to work with as a result. The ethene complex shows IR absorptions at 2083 and 2049 cm^{-1} (CO stretching) and at 1527 cm^{-1} (CC) as expected. The Fe$(CO)_2$Cp group is commonly abbreviated as "Fp", and this will be used throughout our discussion. The starting material for the preparation of alkene-Fp$^+$ complexes is the red crystalline dimer [Cp$(CO)_2$Fe]$_2$ (**2.9**), which is prepared by reaction of cyclopentadiene with iron pentacarbonyl. The dimer can be converted into [CpFe$(CO)_2$]$^-$Na$^+$ by reduction with sodium/mercury amalgam, and used *in situ* as a nucleophile, or it can be oxidatively cleaved by bromine to give CpFe$(CO)_2$Br.

2 CpFe$(CO)_2$Br ←Br_2— [Cp(CO)Fe(μ-CO)$_2$Fe(CO)Cp] **2.9** —Na/Hg→ 2 [CpFe$(CO)_2$]Na (2.8)

Reaction of CpFe$(CO)_2$Br with an alkene in the presence of a suitable Lewis acid, e.g., $AlBr_3$, leads to the desired alkene-Fp cation. This method is, of course, limited to the use of alkenes that are stable to Lewis acids (equation 2.9). Usually the tetrabromoaluminates are converted into hexafluorophosphates by treatment with aqueous ammonium hexafluorophosphate, since these derivatives have a better shelf life and are easier to handle. This method provides access to the isobutene-Fp cation **2.11**, which may also be used in the preparation of other alkene-Fp derivatives by olefin exchange (equations 2.10 and 2.11). The exchange process is facilitated by the loss of gaseous isobutene from the reaction milieu and by a preference for the formation of sterically less encumbered alkene complexes, thereby allowing for selectivity.

CH_2=CRR' —CpFe$(CO)_2$Br, $AlBr_3$→ (CH_2=CRR')–Fp$^+$ $AlBr_4^-$ (or BF_4^-, see preparation 2.4) (2.9)

2.10
2.11 R = R' = Me

2.11 → Fp+ AlF4− 2.12 (65%) (2.10)

2.11 → Fp+ AlF4− 2.13 (76%) (2.11)

The anion $[CpFe(CO)_2]^-$ can be used in a number of indirect ways to gain access to alkene-Fp complexes. Reaction with an alkyl halide followed by hydride abstraction using the triphenylmethyl (trityl) cation (equation 2.12), with an allylic halide followed by protonation (equation 2.13 and Preparation 2.4) or with an epoxide followed by acid-promoted dehydration (equation 2.14) all provide convenient ways to prepare the desired complexes. The latter route has been found particularly useful for the preparation of Fp complexes of α,β-unsaturated ketones (equation 2.15). It should be mentioned here that reaction of allyl-Fp complexes such as **2.14** with electrophiles other than proton leads to the formation of substituted alkene-Fp complexes, but discussion of this will be deferred until Chapter 3.

$CpFe(CO)_2Na \xrightarrow{RCH_2CH_2I} FpCH_2CH_2R \xrightarrow{Ph_3CBF_4}$ R-alkene–Fp^+ BF_4^- (2.12)

$CpFe(CO)_2Na \xrightarrow{CH_2=CHCH_2Cl}$ allyl-Fp **2.14** $\xrightarrow{HBF_4}$ R-alkene–Fp^+ BF_4^- **2.15** (2.13)

$CpFe(CO)_2Na$ 1) ethylene oxide, 2) H_2O → HO–CH2CH2–Fp **2.16** $\xrightarrow{HBF_4}$ ethylene–Fp^+ BF_4^- **2.17** (2.14)

Me–CO–epoxide 1) $CpFe(CO)_2Na$, 2) HBF_4, Et_2O, −78 °C → Me–CO–CH=CH2–Fp^+ BF_4^- **2.18** (2.15)

As one might guess, the complexed double bond in these molecules is completely resistant to attack by electrophiles. Consequently, Fp can be used as an alkene protecting group. It is stable to bromination conditions, hydrogenation and acetoxymercuration, but is easily removed by treatment with sodium iodide in acetone or by warming with acetonitrile (*neutral* reaction conditions). In principle the Fp iodide produced by the first method can be recycled, although this is seldom done in practice. An example of the use of the Fp protecting group is shown in Scheme 2.2.

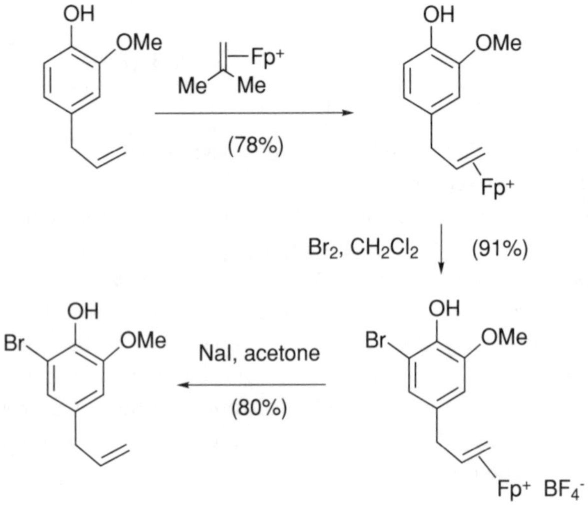

OH OMe — Fp^+ Me Me — (78%) → OH OMe Fp^+

Br_2, CH_2Cl_2 ↓ (91%)

Br OH OMe ← NaI, acetone (80%) — Br OH OMe Fp^+ BF_4^-

SCHEME 2.2. Use of Fp as alkene protecting group.

On the other hand, the alkene is strongly activated toward nucleophile addition, leading to a number of methods for carbon–carbon bond formation. Enolate equivalents, such as enamines and enolsilanes, are particularly useful in these systems. Although CC bond formation cannot be effected using alkyllithium and Grignard reagents, owing to competing electron transfer reactions, organocuprates react satisfactorily to give alkylation products. A large number of nucleophile additions show pronounced regioselectivity, usually occurring at the more highly substituted carbon, presumably as a result of the greater positive charge density at this position. However, the regiocontrol is not always good enough to be of real value in organic synthesis. Some examples of typical nucleophile additions and subsequent transformations are shown in equations 2.16–2.22. Note that the addition is completely stereoselective, *anti* to the Fp group (equation 2.19).

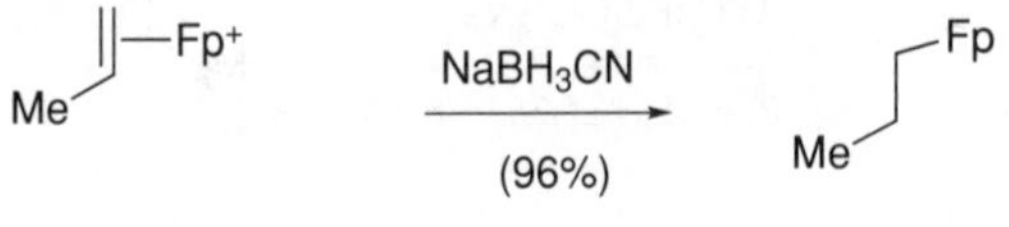

Me —Fp^+ — $NaBH_3CN$ (96%) → Me Fp (2.16)

Fp⁺ → (LiCH($CO_2Me)_2$) → Fp / Ph / CH($CO_2Me)_2$ (2.17)

Fp⁺ / Me → (LiCH($CO_2Me)_2$) → Fp / Me / CH($CO_2Me)_2$ + CH($CO_2Me)_2$ / Me / Fp (2:1) (2.18)

Fp⁺ → (LiCH($CO_2Me)_2$, (82%)) → Fp / CH($CO_2Me)_2$ 2.19)

NR_2 → (Fp⁺) → O / Fp (2.20)

OLi → (**2.18**) → O / O / Fp **2.19** → (Al_2O_3, CH_2Cl_2) → O **2.20** (76%) (2.21)

$OSiMe_3$ / Me **2.21** → (1) **2.18** 2) Al_2O_3) → O / Me **2.22** (2.22)

The enone complex **2.18** undergoes the equivalent of Michael addition, and is highly activated toward this mode of attack. The initial product (**2.19**) undergoes demetallation and intramolecular aldol reaction to give bicyclic molecules, such as **2.20** or **2.22**, the overall reaction corresponding to a Robinson annulation. The clean reaction of the enolsilane **2.21** at the methylated position without loss of regiochemical integrity is particularly interesting, since it avoids some of the problems associated with Lewis acid-catalyzed addition of these nucleophiles to enones.

Heteroatom nucleophiles also add to alkene-Fp complexes (equations 2.23–2.28). In the case of primary amines, further conversion into β-lactams can be effected *via* oxidation of the metal, which induces CO insertion followed by reductive elimination (equations 2.27 and 2.28). This procedure is interesing in view of the importance of β-lactam antibiotics, but it has not seen full exploitation.

$$\text{CH}_2{=}\text{CH}_2{-}\text{Fp}^+\ \text{BF}_4^- \xrightarrow{\text{PR}_3} \text{Fp}\text{CH}_2\text{CH}_2\overset{+}{\text{P}}\text{R}_3\ \text{BF}_4^- \tag{2.23}$$

$$\text{CH}_2{=}\text{CH}_2{-}\text{Fp}^+\ \text{BF}_4^- \xrightarrow{\text{t-BuSH, Na}_2\text{CO}_3} \text{Fp}\text{CH}_2\text{CH}_2\text{SBu}^t \tag{2.24}$$

$$\text{CH}_2{=}\text{CH}_2{-}\text{Fp}^+\ \text{BF}_4^- \xrightarrow{\text{MeOH, Na}_2\text{CO}_3} \text{Fp}\text{CH}_2\text{CH}_2\text{OMe} \tag{2.25}$$

$$\text{CH}_2{=}\text{CH}_2{-}\text{Fp}^+\ \text{BF}_4^- \xrightarrow[\text{(89\%)}]{\text{PhCH}_2\text{NH}_2} \text{Fp}\text{CH}_2\text{CH}_2\overset{+}{\text{N}}\text{H}_2\text{CH}_2\text{Ph}\ \text{BF}_4^- \tag{2.26}$$

$$\text{FpCH}_2\text{CH(Me)}\overset{+}{\text{N}}\text{H}_2\text{CH}_2\text{Ph}\ \text{BF}_4^- \xrightarrow{\text{Cl}_2} \left[\overset{\bullet+}{\text{Fp}}\text{CH}_2\text{CH(Me)}\overset{+}{\text{N}}\text{H}_2\text{CH}_2\text{Ph}\right] \xrightarrow{\text{Me}_3\text{N}} \left[\overset{\bullet+}{\text{Fp}}\text{CH}_2\text{CH(Me)NHCH}_2\text{Ph}\right] \longrightarrow \text{Me-substituted } N\text{-CH}_2\text{Ph } \beta\text{-lactam} \tag{2.27}$$

$$\text{H}_3\text{N}^+(\text{CH}_2)_3\text{CH}{=}\text{CH}_2{-}\text{Fp}^+\ \text{BF}_4^- \xrightarrow[\text{3) Base}]{\text{1) Base; 2) Ag}_2\text{O}} \text{bicyclic } \beta\text{-lactam} \tag{2.28}$$

Complexes of vinyl ethers are readily prepared via α-bromoacetals, and undergo completely regioselective nucleophile additon to the alkoxy-substituted carbon[6,7]. Treatment of the product with tetrafluoroboric acid gives the substituted alkene-Fp complex which can be demetallated to give the substituted alkene. The (vinyl ether)-Fp complexes therefore provide vinyl cation equivalents for reaction with enolate nucleophiles (equations 2.29 and 2.30).

EtO OEt Br —1) NaFe(CO)$_2$Cp 2) HBF$_4$→ EtO —Fp$^+$ BF$_4^-$ **2.23** (2.29)

Me —LDA; EtO Me —Fp$^+$ BF$_4^-$→ O Me OEt Fp Me (93%) —1) HBF$_4$ 2) NaBr→ O CH$_2$ Me Me (86%) (2.30)

The reactions of (vinyl ether)-Fp complexes with enolates are often diastereoselective (equation 2.31), and if these chiral complexes could be produced in optically pure form, a new avenue for asymmetric synthesis would be opened. Progress on this has been made using the dialkoxyalkene complex, **2.24**, which is prepared in 95% yield by reaction of *cis*-1,2-dimethoxyethene with isobutene-FpBF$_4$ (equation 2.32)[8,9]. This complex and the diethoxy derivative, **2.25** can be used for double nucleophile additions (equation 2.33) and extrapolation of this behavior to the use of (*R,R*)-2,3-butanediol allows access to the optically pure complex **2.27** (equation 2.34).

OLi —MeO —Fp$^+$ BF$_4^-$→ O H OMe Fp (2.31)

MeO OMe —Fp$^+$ BF$_4^-$→ Fp$^+$ BF$_4^-$ MeO OMe **2.24** (95%) —EtOH, rt→ Fp$^+$ BF$_4^-$ EtO OEt **2.25** (100%) (2.32)

2.25 —Nucleophile→ OEt Fp R^1 OEt —1) HBF$_4$ 2) Nucleophile 3) HBF$_4$→ Fp$^+$ BF$_4^-$ R^2 R^1 **2.26** (2.33)

2.24 + (2R,3R)-butanediol (Me, Me, HO, OH) ⟶ **2.27** (Fp⁺ BF_4^-) (2.34)

Reactions of **2.27** with nucleophiles are completely stereoselective as shown in equation 2.35. Treatment of the products, e.g., R=H or Me, with trimethylsilyl triflate (CH_2Cl_2, Et_2O, $-78°C$, 10 min) produces the substituted vinyl ether-Fp complexes in diastereomerically and enantiomerically pure form (equation 2.36). However, the utility of these compounds is somewhat limited by their configurational instability, since racemization occurs at room temperature over a period of *ca* 19 h (equation 2.37). Consequently, their application in asymmetric synthesis requires that they be generated and used immediately.

2.27 —Nucleophile→ **2.28** (2.35)

R = H (49%)
R = Me (40%)
R = Ph (56%)
R = 2-oxocyclohexyl (60%)
R = CN (98%)
R = Et (65%)

2.28 (R = H or Me) —Me_3SiOTf→ **2.29** (R = H or Me) (Fp⁺ OTf⁻; Me, Me, $OSiMe_3$) (2.36)

Fp⁺ (OR*, H) ⇌ Fp (+OR*, H) ⇌ Fp (H, OR*+) ⇌ Fp⁺ (H, OR*) (2.37)

Treatment of **2.29** with methanol allows the preparation of optically pure complexes **2.30**, with recovery of the chiral auxiliary, and reaction of **2.30** with nucleophiles, followed by acid treatment, leads to configurationally stable alkene-Fp complexes **2.31**. This shows the importance of substitution in determining the ease of racemization of these complexes *via* a mechanism such as that shown in equation 2.37.

2.29 (R = Me) —MeOH→ **2.30** (78%) —1) $NaBH_4$, NaOMe, MeOH, −78 °C; 2) HBF_4, Et_2O, −78 °C→ **2.31** (69%) (2.38)

Closely related to the above alkene-Fp complexes are the alkyne-Fp and allene-Fp derivatives. The reaction shown in equation 2.10 indicates that alkene-Fp complexes can be formed selectively in the presence of an alkyne. However, in the absence of such competition, alkyne complexes can be readily prepared[10,11]. Alternatively, the triphenylphosphine- or triphenylphosphite-substituted complexes can be prepared by reaction of the alkyne with $CpFe(CO)(PR_3)I$ in the presence of silver tetrafluoroborate (equation 2.39)[12,13]. The latter complexes undergo reaction with a range of carbon and heteroatom nucleophiles, to give σ-alkenyl-$Fe(CO)(PR_3)Cp$ complexes (equations 2.40–2.43).

$CpFe(CO)(PR_3)I$ —1) $AgBF_4$, CH_2Cl_2; 2) 2-Butyne→ **2.32** (70 – 90%) (2.39)

2.32 —PhSNa, 80 – 86%→ **2.33** ([Fe] = $Cp(CO)(PR_3)Fe$) (2.40)

2.32 —$R_2Cu(CN)Li_2$, 62 – 87%→ **2.34** R = Me; Ph; CH_2=CH; Me−C≡C (2.41)

$$\mathbf{2.32} \xrightarrow[48-87\%]{CN^-} \text{(Me)(CN)C=C([Fe])(Me)}\ \mathbf{2.35} \qquad (2.42)$$

$$\mathbf{2.32} \xrightarrow[80-86\%]{NaCH(CO_2Et)_2} \text{(Me)(CH(CO_2Et)_2)C=C([Fe])(Me)}\ \mathbf{2.36} \qquad (2.43)$$

In most cases, nucleophile additions to complexes of unsymmetrically disubstituted alkynes lead to mixtures of regioisomers. The methoxymethyl-substituted derivative, **2.37**, however, reacts with nucleophiles regioselectively (equation 2.44)[14]. Controlled oxidation of the vinyl-Fe complexes that result from nucleophile addition, in the presence of carbon monoxide, leads to CO insertion to give the unsaturated acyliron complexes **2.40** (equation 2.45). Treatment of these compounds with excess of oxidizing agent (3 equivalents) in an alcohol as solvent gives the ester derivatives **2.41** (equation 2.46), and this provides a method for stereocontrolled construction of tetrasubstituted alkenes (equation 2.47).

$$\text{MeC}\equiv\text{CCH}_2\text{OMe}\text{-}\overset{+}{Fe}(PR_3)(CO)Cp\ BF_4^- \xrightarrow{Ph_2Cu(CN)Li_2} \text{(Me)(Ph)C=C([Fe])(CH_2OMe)}\ \mathbf{2.38}\ (80\%) \qquad (2.44)$$

$$\text{(R}^3\text{)(R}^2\text{)C=C([Fe])(R}^1\text{)}\ \mathbf{2.39} \xrightarrow[CO,\ CH_2Cl_2]{[Cp_2Fe]BF_4\ \textit{or}\ Ce(IV),\ 15\ mol\%,} \text{(R}^3\text{)(R}^2\text{)C=C(C(=O)[Fe])(R}^1\text{)}\ \mathbf{2.40} \qquad (2.45)$$

$$\mathbf{2.39}\ \text{or}\ \mathbf{2.40} \xrightarrow[CO,\ ROH]{\text{Excess [O] as equation 2.45}} \text{(R}^3\text{)(R}^2\text{)C=C(C(=O)OR)(R}^1\text{)}\ \mathbf{2.41} \qquad (2.46)$$

As above, using MeOH

2.38 → **2.42** (2.47)

Reactions of allene-Fe(CO)LCp complexes with nucleophiles also give vinyl-[Fe] complexes which can, in principle, be converted into unsaturated esters as outlined for the above compounds (equation 2.48)[15–17].

Nucleophile

(2.48)

2.3 EXPERIMENTAL PROCEDURES

Preparation of tetracarbonyl(maleic anhydride)iron **(2.5)** *and tetracarbonyl(methyl acrylate) iron* **(2.4)**. *(Translated and modified from the German text*[2]*)*

Equimolar amounts of maleic anhydride (0.05 moles, 4.90 g freshly sublimed) and nonacarbonyldiiron (18.20 g) are suspended in absolute benzene (50 ml) under argon atmosphere and stirred for 4 h at 45°C. As the reaction proceeds, a yellow precipitate of complex forms, and this is filtered, washed with a small volume of benzene and dried under vacuum (yield: 11.16 g). From the yellow filtrate, after removal of the solvent and of the iron pentacarbonyl that is also produced during the reaction *(CAUTION: Pentacarbonyliron is toxic: see p. 90 for methods of removal and destruction)*, an additional 0.73 g of complex can be obtained (total crude yield = 89%). The crude product (3.00 g) is dissolved in 100 ml of acetone. The solution is filtered, concentrated and set aside at 0°C. The resulting crystals are filtered off and dried under vacuum to give analytically pure compound (2.34 g, 70%), m.p. 148°C. IR (KBr) 2087, 2055, 2045, 2032, 2017, 1824, 1746 cm^{-1}. Tetracarbonyl(methyl acrylate)iron (**2.4**) is prepared in a similar fashion, but it is soluble in benzene and is obtained from the filtered reaction solution after removal of the solvent and iron pentacarbonyl with a water aspirator. The product is purified by vacuum sublimation (yield: 71%), m.p. 28–28.5°C. IR (heptane) 2101, 2036, 2020, 1996, 1713 cm^{-1}.

Enolate addition to tetracarbonyl(methyl acrylate)iron (equation 2.7)

Complete experimental details are not given in the literature, and the following description is an adaptation of the procedure described in ref. [3]. To a stirred solution of tetracarbonyl(methyl acrylate)iron (**2.4**, 254 mg, 1 mmol, from the above preparation) in THF (5 ml) under argon at 0°C, is added dropwise a solution of diethyl sodiomalonate (364 mg, 2 mmol, prepared from NaH and di*ethyl* malonate as described on p. 157) in THF (5 ml). The mixture is stirred at 0°C for 20 h, after which time trifluoroacetic acid (0.5 ml) is added. After stirring for approximately 30 min at room temperature, the reaction mixture is poured into water (50 ml) and the product is extracted with ether (3 × 10 ml). The combined ether extracts are washed with water (2 × 10 ml), aqueous sodium carbonate (10 ml), dried ($MgSO_4$) and evaporated to give the crude product which can be purified by preparative TLC (silica gel, 20% EtOAc in hexane). Yield: 92%

Preparation of cyclopentadienyldicarbonyliron dimer $[CpFe(CO)_2]_2$ (**2.9**)

The procedure is taken from ref. [18]. Pentacarbonyliron (10 ml) and an excess of dicyclopentadiene (10 ml) are heated under reflux and nitrogen atmosphere, with exclusion of light, for approximately 40 h. Volatiles are removed under vacuum *(see p. 90 for precautions on dealing with unreacted pentacarbonyliron),* and chloroform (50 ml) is added. Nonacarbonyldiiron and other insoluble material is removed by centrifugation and the product is crystallized out by cooling the solution to −78°C (solid CO_2–acetone bath). The crystals are centrifuged out and recrystallized from chloroform to give **2.9** as dark reddish-purple crystals in around 30% yield based on pentacarbonyliron. M.p. 194°C, d. (evacuated sealed tube). The complex is readily soluble in alcohol, chloroform, and pyridine, less soluble in carbon tetrachloride and carbon disulfide and is only sparingly soluble in petroleum ether (red solutions in all solvents). It is insoluble in and stable to water, is stable to air and light, but solutions are less stable. It is converted into ferrocene in *ca* 75% yield on heating to 210°C.

Preparation of dicarbonyl(cyclopentadienyl)isobuteneiron tetrafluoroborate (**2.11**)

The procedure is taken from ref. [19]. This complex is useful as a reagent for transfer of the Fp cation to other alkenes.

(i) *Preparation of methallyl-Fp precursor: CAUTION: Care should be exercised in the preparation of the sodium/mercury amalgam because the initial reaction is highly exothermic. This and all subsequent operations should be carried out in a well-ventilated fume hood.* A 500-ml three-neck flask with a stopcock at the bottom is fitted with a nitrogen inlet and an overhead stirrer with a Teflon paddle. The flask is flame-dried with nitrogen flushing and then 30 ml of mercury is introduced. A pan should be placed under the flask in case of breakage. The mercury is stirred vigorously as sodium metal (4.5 g, 0.196 mol), cut into small pieces, is slowly added. (*CAUTION: The amalgamation of sodium is highly exothermic. Small pieces of sodium must be added to mercury behind a shield.)* The flask is capped with a rubber septum, and the resulting hot amalgam is allowed to reach room temperature. Tetrahydrofuran (200 ml, predried over KOH and then freshly distilled under a nitrogen atmosphere from sodium benzophenone ketyl) is added via a 10-gauge cannula (Hamilton) inserted through rubber septa capping both delivery and receiver vessels. Transfer is made by positive nitrogen pressure applied through a hypodermic needle, while a second needle in the receiver is used as a vent. Dicarbonyl(η^5-cyclopentadienyl)iron dimer (18 g, 0.051 mol, see p. 32) is added in a single portion, and the mixture is vigorously stirred at room temperature for 30–45 min.

The progress of the reaction can be monitored by following changes in the carbonyl region of the IR spectrum of the solution, using carefully dried sodium chloride liquid sample cells filled by syringe under nitrogen. Carbonyl absorption bands of the dimer $[Fe(CO)_2(\eta^5\text{-}C_5H_5)]_2$ at 1995, 1950, and 1780 cm^{-1} are replaced by those of the salt, which exhibits strong absorption bands at 1877 and 1806 cm^{-1} due to the tight ion pair as well as weaker absorptions at 1862, 1786, and 1770 cm^{-1} due to solvent-separated and carbonyl-bridged ion pairs. The amalgam is drained, and the amber–red solution of sodium dicarbonyl(η^5-cyclopentadienyl)ferrate(1-) is used without further purification.

The solution is cooled in an ice bath and is stirred rapidly while 3-chloro-2-methyl-1-propene (9.65 g, 0.107 mol) is added over a period of 5 min. The reaction can be followed by observing the changes in the IR spectrum of the solution. The carbonyl absorption bands characteristic of the anion are replaced by those typical of the product at 1998 and 1950 cm^{-1}. On completion of the addition of 3-chloro-2-methyl-1- propene, stirring at 0°C is continued for 1 h to ensure completion of the reaction. The product is isolated and purified by removing solvent under reduced pressure, followed by chromatography of the residue on 300 g of neutral activity alumina (III). The column is made up in anhydrous diethyl ether, and after dissolving the crude product in petroleum ether, elution under N_2 is carried out with this solvent. The product may be further purified by short-path distillation at pressures less than 10^{-4} mm (pot temperature less than 40°C). It is then sufficiently pure to be stored at −20°C for prolonged periods without decomposition. The yield of dark amber oil is typically 19–20 g (80–90%). IR (CH_2Cl_2): 2003 and 1945 cm^{-1}. ^1H NMR ($CDCl_3$): 4.63 (s, 5H, Cp), 4.47 (m, 2H, CH_2=), 2.11(s, 2H, CH_2), and 1.77 (s, 3H, CH_3).

(ii) *Conversion of methallyl-Fp to isobutene-Fp(+1) BF_4^-*: The above product is placed in a dry, 1-l, single-neck, round-bottom flask with a side arm. The flask is flushed with nitrogen and fitted with a magnetic stirring bar and a rubber septum. Anhydrous diethyl ether (300 ml) is degassed by purging for several minutes with a stream of dry nitrogen, using a gas dispersion tube, and is then transferred to the 1-l flask by cannula. The solution is cooled to 0°C in an ice bath, and 48% aqueous tetrafluoroboric acid (17 ml, 0.12 mol) is added slowly by syringe while the solution is stirred vigorously. Manual shaking may be necessary at the end to ensure mixing of the reactants. A yellow–orange precipitate forms immediately. The septum is removed, and the mixture is transferred by a 2.5-mm cannula to a Schlenk tube fitted with a coarse-porosity sintered glass filter. The product is washed with anhydrous ether until the washings are colorless and is then dried by passing a stream of dry nitrogen through the Schlenk tube. The crude product may be purified as follows. The Schlenk tube receiver is replaced by a 500-ml round-bottom flask with a magnetic stirring bar. The Schlenk tube outlet stopcock is closed, and the crude salt is taken up in 30 ml of methylene chloride (previously dried over 4 Å molecular sieves and then deoxygenated by nitrogen purge). The stopcock is then opened, and the resulting cherry red solution is vacuum-filtered into the round-bottom flask. The process is repeated several times with smaller portions of methylene chloride until the washings are colorless. The Schlenk tube is then replaced by a rubber septum, and the dichloromethane solution is cooled in an ice bath and stirred vigorously as anhydrous diethyl ether (250 ml) is added over a period of 5 min. The resulting golden yellow solid is transferred as before to a Schlenk tube with a medium-porosity filter. The filter cake is washed several times with small portions of diethyl ether, dried under nitrogen in the Schlenk tube, and finally dried under vacuum. The yield of yellow crystalline dicarbonyl(η^5-cyclopentadienyl)(η^2-2-methyl-1-propene)iron tetrafluoroborate (**2.11**) is 25–28 g (78–88%). IR (CH_3NO_2): 2030, 2070 cm^{-1}. ^1H NMR (CD_3NO_2): 5.64 (s,5H, Cp), 3.91 (s, 2H, CH_2=), 1.96 (s, 6H, CH_3). The product may be stored indefinitely under nitrogen at −20°C without decomposition. It is soluble in methylene chloride, acetone, and nitromethane but insoluble in hydrocarbons and in diethyl ether.

General procedure for the preparation of alkene-Fp cations by ligand exchange with isobutene-Fp tetrafluoroborate

The procedure is described for the preparation of the η^2 complex of 3-methoxypropene. Other alkene complexes are prepared by modification of this method, by adjusting the reaction time to give optimum yield[20]. A solution of the isobutene-Fp salt (**2.11**, 960 mg, 3.0 mmol) in 1,2-dichloroethane (15 ml) is heated to 70°C under a reflux condenser and nitrogen bubbler, and a solution of 3-methoxypropene (2.5 ml, 30 mmol) in 1,2-dichloroethane (5 ml) is added rapidly via syringe. The reaction mixture is stirred at 70°C for 3 min and then cooled to room temperature. Ether is added until no more precipitate forms, and the insoluble product is removed by filtration and recrystallized from methylene chloride/ether, to give dicarbonyl(cyclopentadienyl)(3-methoxypropene)iron tetrafluoroborate (710 mg, 71% yield), dec 100°C. IR (KBr) 2049, 2000 cm^{-1}. ^1H NMR (CD_3NO_2) δ 5.68 (5H, s, Cp), 5.3 (1H, m, Vinyl), 4.04 (1H, d, J 8 Hz, *cis* = CH), 3.98 (2H, d, J 4 Hz, CH_2), 3.56 (1H, d, J 15 Hz, *trans* = CH).

General procedure for the preparation of alkene-Fp cations from epoxides (equations 2.14 and 2.15).

The method is taken from ref. [21]. The epoxide is added slowly to an equimolar amount of NaFp (see pp. 32–33) in THF solution at 0°C. After the addition is complete, the stirred solution is brought to room temperature and the reaction is continued for 30 min to 12 h, depending on the epoxide that is used. For example, ethylene oxide, propene oxide, 1-butene oxide all require only 30 min, *cis*- and *trans*-2-butene oxide both require 12 h, while cyclohexene oxide requires 4 h. The solution usually turns dark green when the reaction is complete. The stirred solution is cooled to 0°C (or −78°C if the product cation is unstable) and 48% aqueous tetrafluoroboric acid is added slowly (70% aqueous hexafluorophosphoric acid can be used to obtain alkene-Fp hexafluorophosphates). Ether is added until precipitation is complete, and the product is removed by filtration, washed with ether and dried under vacuum. The salts can be recrystallized from methylene chloride–acetone or nitromethane–ether, and are generally obtained as yellow air-stable solids that decompose on heating. Yields: ethene-$FpBF_4$, 90%; propene-$FpBF_4$, 91%; styrene-$FpBF_4$, 65%; (*cis*-2-butene)$FpBF_4$, 64%; (*trans*-2-butene)$FpBF_4$, 50%; cyclohexene-$FpBF_4$, 60%.

References

1. H.D. Murdoch and E. Weiss, *Helv. Chim. Acta*, **46**, 1588 (1963).
2. E. Weiss, K. Stark, J.E. Lancaster and H.D. Murdoch, *Helv. Chim. Acta* **46**, 288 (1963).
3. B.W. Roberts and J. Wong, *J. Chem. Soc. Chem. Commun.* 20 (1977).
4. B.W. Roberts, M. Ross and J. Wong, *J. Chem. Soc. Chem. Commun.* 428 (1980).
5. M. Rosenblum, *Acc. Chem. Res.* **7**, 122 (1974).
6. T.C.T. Chang, M. Rosenblum and S.B. Samuels, *J. Am. Chem. Soc.* **102**, 5930 (1980).
7. T.C.T. Chang and M. Rosenblum, *J. Org. Chem.* **46**, 4103 (1981).
8. M. Marsi and M. Rosenblum, *J. Am. Chem. Soc.* **106**, 7264 (1984).
9. M. Rosenblum, M.M. Turnbull and B.M. Foxman, *Organometallics* **5**, 1062 (1986).
10. D.J. Bates, M. Rosenblum and S.B. Samuels, *J. Organomet. Chem.* **209**, C55 (1981).
11. J. Benaim and A. L'Honore, *J. Organomet. Chem.* **202**, C53 (1981).
12. D.L. Reger, K.A. Belmore, E. Mintz and P.J. McElligot, *Organometallics* **3**, 134 (1984).
13. D.L. Reger, S.A. Klaeren and L. Lebioda, *Organometallics* **5**, 1072 (1986).
14. D.L. Reger, S.A. Klaeren, J.E. Babin and R.E. Adams, *Organometallics* **7**, 181 (1988).
15. D.W. Lichtenberg and A. Wojcicki, *J. Organomet. Chem.* **94**, 311(1975).
16. P. Lennon, A.M. Rosan and M. Rosenblum, *J. Am. Chem. Soc.* **99**, 8426 (1977).
17. D.W. Lichtenberg and A. Wojcicki, *J. Am. Chem. Soc.* **94**, 8271 (1972).
18. T.S. Piper, F.A. Cotton and G. Wilkinson, *J. Inorg. Nucl. Chem.* **1**, 165 (1955).
19. M. Rosenblum, W.P. Giering and S-B. Samuels, *Inorg. Syn.* **28**, 207 (1990).
20. A. Cutler, D. Entholt, P. Lennon, K. Nicholas, D.F. Marten, M. Madhavarao, S. Raghu, A. Rosan and M. Rosenblum, *J. Am. Chem. Soc.* **97**, 3149 (1975).
21. A. Cutler, D. Entholt, W.P. Giering, P. Lennon, S. Raghu, A. Rosen, M. Rosenblum, J. Tancrede and D. Wells, *J. Am. Chem. Soc.* **98**, 3495 (1976).

–3–

η^1 -and η^3-Allyliron Complexes. η^1-Acyliron Complexes. Carbene Complexes

3.1 ALLYL-FP COMPLEXES

Reaction of the nucleophilic complex $[Cp(CO)_2Fe]^-Na^+$ with allylic halides gives σ-allyl-Fp complexes in good yields ($Fp{=}Fe(CO)_2Cp$, see Chapter 2). Alternatively, deprotonation of η^2-alkene-Fp complexes, discussed in Chapter 2 that have allylic hydrogens can be used to generate various substituted derivatives (equations 3.1–3.3).

$$[CpFe(CO)_2]_2 \xrightarrow{Na/Hg} 2\ [CpFe(CO)_2]Na \xrightarrow{CH_2{=}CHCH_2X} CH_2{=}CHCH_2Fp \quad \mathbf{3.1} \tag{3.1}$$

$$XCH_2CH{=}CH_2{-}Fp^+ \xrightarrow{Et_3N} XCH{=}CHCH_2Fp \tag{3.2}$$

3.2 X = MeO
3.3 X = Br

$$RCH_2CH{=}CH_2{-}Fp^+ \xrightarrow{Et_3N} RCH{=}CHCH_2Fp \tag{3.3}$$

3.4 R = alkyl

The deprotonation reactions shown in equations 3.2 and 3.3 are quite stereoselective and this has been rationalized on the basis of conformational preferences for the alkene-Fp complexes. Simple alkyl-substituted compounds appear to prefer the conformation **A**, whereas **B** is favored for those complexes having electron-withdrawing heteroatom substituents on the allylic carbon, owing to the stabilizing interaction between the lone pair and a carbonyl ligand as indicated. Stereoelectronic effects dictate removal of the proton *anti* to the Fp group, resulting in *E* double bond for alkyl-substituents, and *Z* double bond for heteroatom substituted compounds.

OC CO Fe H R H H H Base

A

OC CO Fe X H H H H Base

B

The reactions of allyl-Fp complexes with electrophiles and in non-concerted [3+2] cycloaddition reactions have been studied primarily by Rosenblum[1] and Baker[2–5]. In terms of their reactions with electrophiles, they may be thought of in much the same way as allylsilanes or allyltin reagents, since they can lead to products of allylation of the electrophile. They are less reactive than allyl Grignard reagents. As simple allylating reagents they offer no real advantage over the silicon, tin or Grignard counterparts. However, in contrast to these reagents, the initial product from the allyl-Fp/electrophile reaction is an isolated η^2-alkene-Fp cation, which is itself an electrophile. Therefore, if this species can be induced to react with nucleophiles in a stereo- and/or regiocontrolled manner, the allyl-Fp system would provide a powerful intermediate for organic synthesis. So far, very little attention has been paid to this aspect of reactivity, except with the [3+2] cycloaddition reactions to be discussed later. Some examples of reactions of allyl-Fp complexes with simple electrophiles are given in equations 3.4–3.10.

3.1 Fp —[SO_2, $Me_3O^+ BF_4^-$]→ Fp^+–(alkene)–CH_2SO_2Me BF_4^- **3.5** (80%) (3.4)

3.1 —[$Me_3O^+ BF_4^-$]→ Fp^+–(alkene)–Me BF_4^- **3.6** (3.5)

3.1 —[RCOX, $AgSbF_6$]→ Fp^+–(alkene)–COR SbF_6^- **3.7** (3.6)

3.1 → (Ph–dioxolanylium) → Fp^+ … **3.8** (95%) (3.7)

3.1 → 1) Br_2, –78 °C; 2) HPF_6 → Fp^+ … Br PF_6^- **3.9** (87%) (3.8)

3.1 → $HgCl_2$ → Fp^+ … HgCl **3.10** (3.9)

Fp … Me **3.11** → HPF_6 → Fp^+ … Me PF_6^- **3.12** (3.10)

Treatment of allyl-Fp complexes with electron deficient alkenes results in a tandem electrophile/nucleophile addition sequence to give products corresponding to overall [3+2] cycloaddition. This is illustrated schematically in equation 3.11, followed by a number of typical examples in equations 3.12–3.19. The net result is the formation of a five-membered ring, which contains a σ-bound Fp group. The stereochemical leakage shown in equation 3.17 is consistent with a non-concerted mechanism.

Fp … Y=X → Fp^+ … Y^-–X → Fp … Y, X (3.11)

Fp, R → tetracyanoethylene, CH_2Cl_2, 5 min → Fp, R, CN, CN, CN, CN (3.12)

3.13 R = H or Me

Fp; 3.1; CO_2Me; CO_2Me; CO_2Me; CO_2Me; Fp; 3.14 (3.13)

Cl; Cl; O; O; CN; CN; 3.1; O; O; Cl; Cl; CN; CN; Fp; 3.15 (3.14)

Fp; 3.16; CNTs; Fp; H; NTs; O; 3.17 (3.15)

Fp; R; CNTs; Ts; N; O; Fp; R; 3.18 (3.16)

Fp; Ph; 3.19; F_3C; CN; NC; CF_3; CN; CF_3; CN; CF_3; Fp; Ph; 3.20; +; CN; CF_3; CF_3; CN; Fp; Ph; 3.21 (3.17)

Fp; OMe; NC; CO_2Me; MeO_2C; H; CH_2Cl_2, rt; CN; CO_2Me; CO_2Me; Fp; OMe; 3.22 (86%) (3.18)

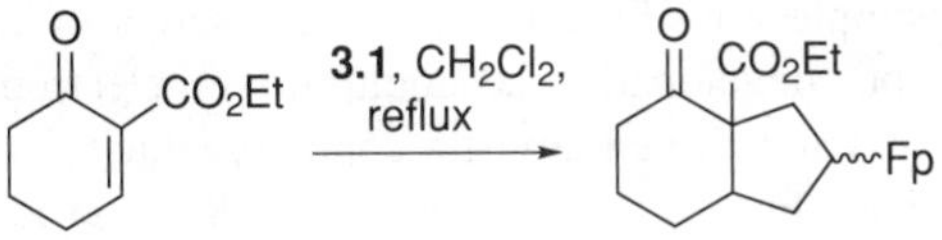

O, CO_2Et — 3.1, CH_2Cl_2, reflux → O, CO_2Et, Fp

3.23 (3.19)

These cycloadditions are quite sensitive to steric hindrance, both in the allyl-Fp reagent and in the alkene partner, very highly substituted derivatives giving no useful products. There are a number of ways to remove the Fp group from the product σ-alkyl complexes. Conversion into the cationic alkene-Fp complex allows decomplexation by treatment with iodide, bromide, or acetonitrile, or by photochemical methods (see also Chapter 2). Alternatively, oxidation with ceric ammonium nitrate in methanol leads to CO insertion followed by demetallation to give a methyl ester. These methods are shown schematically in equations 3.20 and 3.21. An example of the application of this protocol to the synthesis of a sarkomycin intermediate **3.26** is shown in Scheme 3.1.

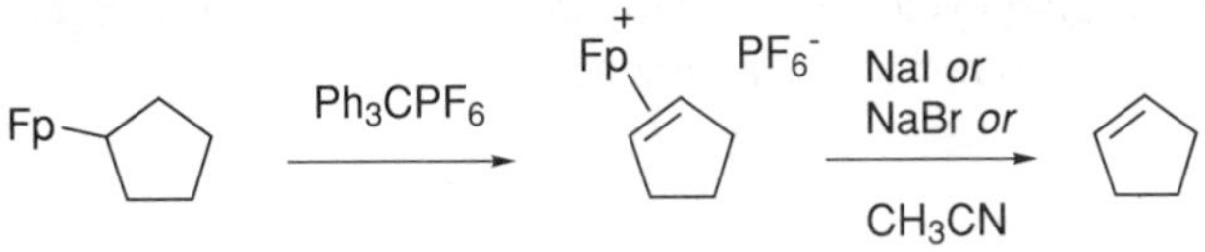

Fp — Ph_3CPF_6 → Fp^+ PF_6^- — NaI *or* NaBr *or* CH_3CN → (3.20)

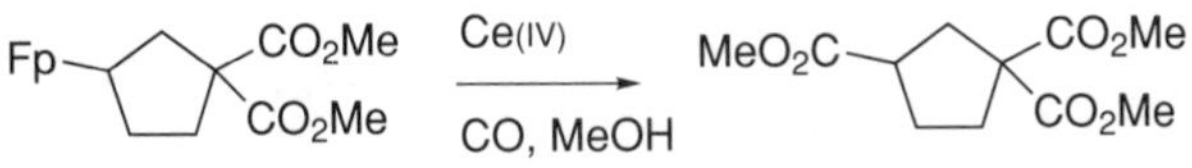

Fp — CO_2Me, CO_2Me — Ce(IV), CO, MeOH → MeO_2C — CO_2Me, CO_2Me (3.21)

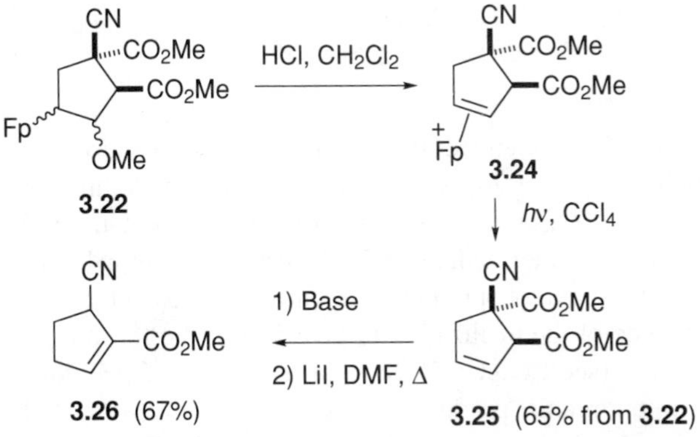

CN, CO_2Me, CO_2Me, Fp, OMe **3.22** — HCl, CH_2Cl_2 → CN, CO_2Me, CO_2Me, Fp^+ **3.24**

3.24 — *hν*, CCl_4 → CN, CO_2Me, CO_2Me **3.25** (65% from **3.22**)

3.25 — 1) Base, 2) LiI, DMF, Δ → CN, CO_2Me **3.26** (67%)

SCHEME 3.1

Starting from a simple allyl-Fp complex **3.1**, a series of carbon–carbon bond constructions can be developed *via* addition of electrophile, followed by deprotonation to give a substituted allyl-Fp complex which is then used in a [3+2] cycloaddition reaction. For example, complex **3.8**, a product of electrophile addition, is deprotonated to give **3.27** (equation 3.22) which reacts with tetracyanoethylene (TCNE) to give **3.28**.

3.8 —Et_3N→ 3.27 —tetracyano-ethylene→ 3.28 (3.22)

Variations on the above cycloadditions include the use of the cyclopropylmethyl complex **3.29** to give six-membered rings[6], and the use of the σ-alkenyl-Fp complex **3.31** to give cyclopentene derivatives, although these reactions have not yet seen application to real synthetic problems.

—NaFp→ —tetracyano-ethylene→ 3.30 (3.23)

—NaFp→ 3.31 (52%) —tetracyano-ethylene→ 3.32 (3.24)

Rosenblum[7] has extended these cycloaddition reactions to give a "double organometallic" approach to hydroazulenes, which are components of a number of biologically active sesquiterpenes. Using the η^5-cycloheptatrienyliron complex **3.33** as electrophile, the intermediate **3.34** is generated, in which the uncomplexed double bond of the cycloheptatriene complex is highly nucleophilic (Scheme 3.2). The driving force of the cyclization to give **3.35** is the formation of the very stable dienyliron cation (see Chapter 5). Several reactions of **3.35** with nucleophiles, and conversion into organic products are shown in Scheme 3.2. Extension of this strategy to the use of the more highly substituted allyl-Fp derivatives **3.40** gives the diastereomeric bicyclic complexes **3.41** and **3.42**.

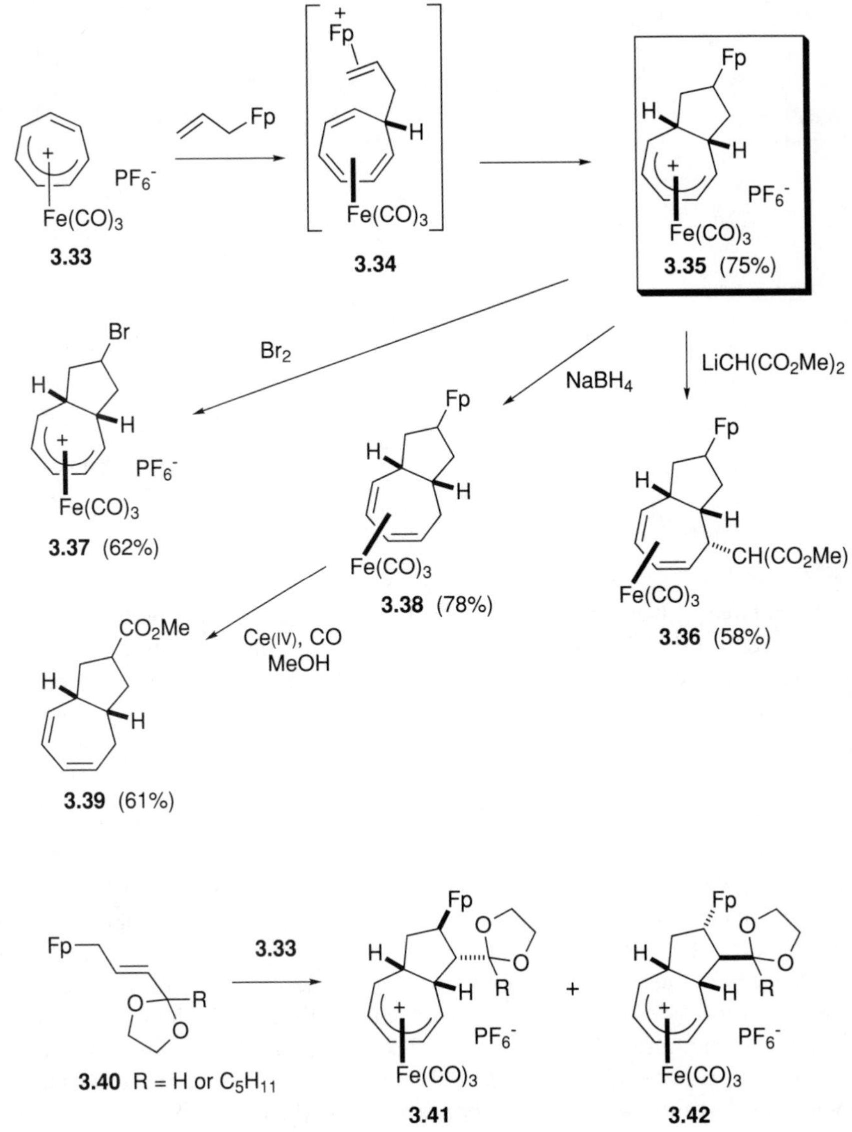

SCHEME 3.2

Even more interesting is the use of complex **3.44**, obtained from the methylcycloheptatrienone complex **3.43** (Scheme 3.3). Reaction of **3.44** with allyl-Fp occurs with complete regioselectivity to give **3.46** (*via* **3.45**)[8,9], and this can be demetallated in the usual way to give the hydroazulene derivative **3.47**, having an angular methyl substituent.

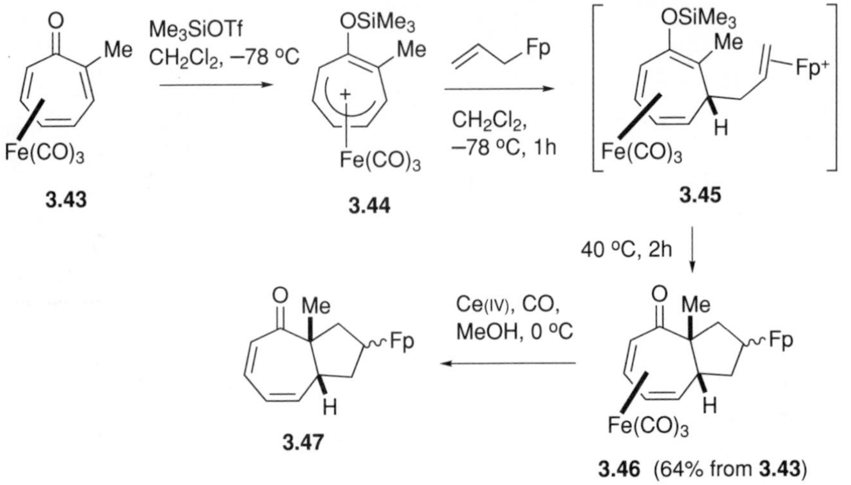

O
Me
Fe(CO)3
3.43
Me3SiOTf
CH2Cl2, −78 °C
OSiMe3
Me
+
Fe(CO)3
3.44
Fp
CH2Cl2,
−78 °C, 1h
OSiMe3
Me
Fp+
H
Fe(CO)3
3.45
40 °C, 2h
O
Me
Fp
H
Fe(CO)3
3.46 (64% from 3.43)
Ce(IV), CO,
MeOH, 0 °C
O
Me
Fp
H
3.47

SCHEME 3.3

In summary, there exists a great deal of potential new methodology for the construction of highly functionalized five-membered ring derivatives using allyl-Fp reagents, but this has not yet been extensively applied in target-oriented synthesis.

3.2 η³-ALLYLIRON COMPLEXES

It is possible to generate these complexes both stoichiometrically and under catalytic conditions. Although they present possibilities for organic synthesis applications that are very similar to the corresponding π-allyl-palladium derivatives[10], they have not seen such extensive development. Consequently, we shall present a brief overview of their chemistry here, in the hope that this will stimulate more research in this area. The parent allyl complex **3.49** can be prepared *via* reaction of allyl iodide with iron pentacarbonyl, to give **3.48** which is then treated with silver tetrafluoroborate under carbon monoxide atmosphere to effect removal of iodide and incorporation of an extra CO ligand (equation 3.25)[11,12].

I
Fe(CO)5, Δ
Fe(CO)3I
3.48
AgBF4
CO
+
Fe(CO)4
BF4-
3.49

(3.25)

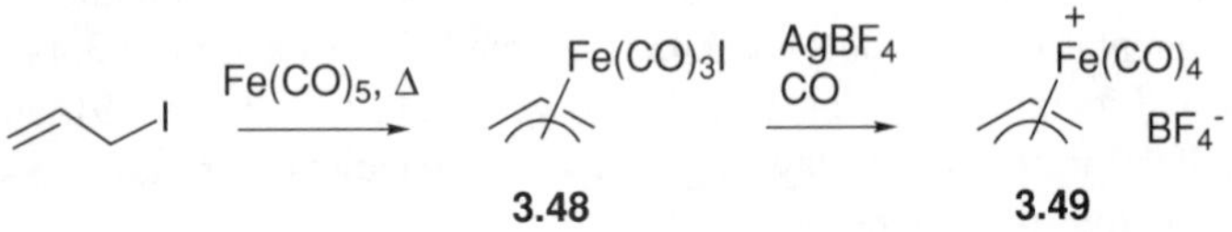

Substituted allyl-$Fe(CO)_4$ cations are best prepared by reaction of a diene-$Fe(CO)_3$ complex with an appropriate electrophile (usually proton) under CO atmosphere[13–18]. The initial product of diene protonation retains the original diene geometry, but when exposed to acid for prolonged periods of time, isomerization to the thermodynamically more stable isomer occurs (e.g. **3.52, 3.53**). Some typical examples are shown in equations 3.26–3.28. Although this method works well for acyclic π-allyl complexes, it has not been possible so far to prepare and isolate the corresponding cyclohexenyl-$Fe(CO)_4$ cations.

$Fe(CO)_3$ — HBF_4, CO → $Fe(CO)_4^+$ BF_4^- (3.26)

3.50 R = Me, R′ = H
3.51 R = R′ = H
3.52 R = H, R′ = Me

3.52 — H^+ → **3.53** (3.27)

3.54 — HBF_4, CO → **3.55** (3.28)

The resulting π-allyl complexes are very reactive toward nucleophiles. The initial products from nucleophile addition are alkene-$Fe(CO)_4$ complexes, which are quite unstable. These usually decompose by air oxidation during the reaction work up, giving substituted alkenes as the isolated products (equations 3.29–3.31).

3.50 — 1) $NaCH(COMe)CO_2Me$; 2) NaOH; 3) Δ ($-CO_2$) → **3.56** (68%) + **3.57** (17%) (3.29)

$$\text{3.50} \xrightarrow{PPh_3} \text{3.58 (74\%)} \quad (3.30)$$

$$\text{3.51} \xrightarrow{R_2Cd} \text{3.59} \quad (3.31)$$

Methodology has been developed[19,22] for the catalytic generation of π-allyliron complexes in the presence of nucleophiles that allows alkylation of allylic acetates (equations 3.32–3.36), in much the same way as the palladium-catalyzed processes[11,12]. Unfortunately, this appears to lack the versatility shown in organopalladium chemistry and has received little attention, despite the lower cost of the iron catalysts. However, optically active allylic carbonates have been alkylated with malonate enolate to give optically active products[20,21,23].

$$\text{CH}_2{=}\text{CHCH}_2\text{OAc} \xrightarrow[NaFe(CO)_3NO]{NaCH(CO_2Et)_2} \text{CH}_2{=}\text{CHCH}_2\text{CH(CO}_2\text{Et)}_2 \text{ (74\%)} \quad (3.32)$$

$$\text{PhCH{=}CHCH}_2\text{OAc} \xrightarrow[NaFe(CO)_3NO]{NaCH(CO_2Et)_2} \text{PhCH{=}CHCH}_2\text{CH(CO}_2\text{Et)}_2 \text{ (91\%)} \quad (3.33)$$

$$\text{CH}_2{=}\text{C(Me)CH}_2\text{OAc} \xrightarrow[NaFe(CO)_3NO]{NaCH(CO_2Et)_2} \text{CH}_2{=}\text{C(Me)CH}_2\text{CH(CO}_2\text{Et)}_2 + \text{MeCH{=}CHCH}_2\text{CH(CO}_2\text{Et)}_2 \text{ (79\% yield; 95:5 ratio)} \quad (3.34)$$

As was mentioned in Chapter 2, vinyl epoxides react with iron carbonyls to give π-allyliron complexes. This reaction has been studied extensively and has been used for the synthesis of β-lactone and β-lactam derivatives[24,25]. The earlier work on π-allyl-ferrelactone complexes was reported by Heck[26], Murdoch[27], Aumann[28,29], and Churchill[30] and their co-workers. Vinyloxiranes are readily available *via* monoepoxidation of dienes. Thermal reaction with iron pentacarbonyl gives diastereomeric mixtures of ferrelactone complexes, but Churchill *et al.*[30] found that the photochemical reaction occurs with complete stereoselectivity (equations 3.35–3.37).

Fe(CO)5, C6H6, hν: **3.60** → **3.61** (93%) (3.35)

Fe(CO)5, C6H6, hν: **3.62** → **3.63** (82%) (3.36)

Fe(CO)5, C6H6, hν: **3.64** → **3.65** (84%) (3.37)

Treatment of the ferrelactones thus obtained with ceric ammonium nitrate results in demetallation with the formation of predominantly β-lactones, except in some cases where δ-lactones are obtained (equations 3.38–3.40). Both pathways have been used to effect the construction of useful synthetic intermediates[24,25]. Alternatively, carbonylation under CO pressure can be used to produce the δ-lactone derivatives.

→ Fe(CO)3; Ce(IV), H2O, EtOH → (3.38)

(3.39)

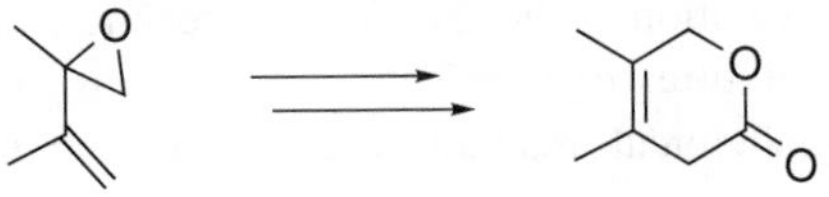

(3.40)

Ferrelactones react with amines to give ferrelactam derivatives, and these may be demetallated to give β-lactams. The significance of this result is obvious, and the method has been used to synthesize thienamycin (**3.66**, Scheme 3.4)[31] and compounds related to nocardicins (Scheme 3.5)[32].

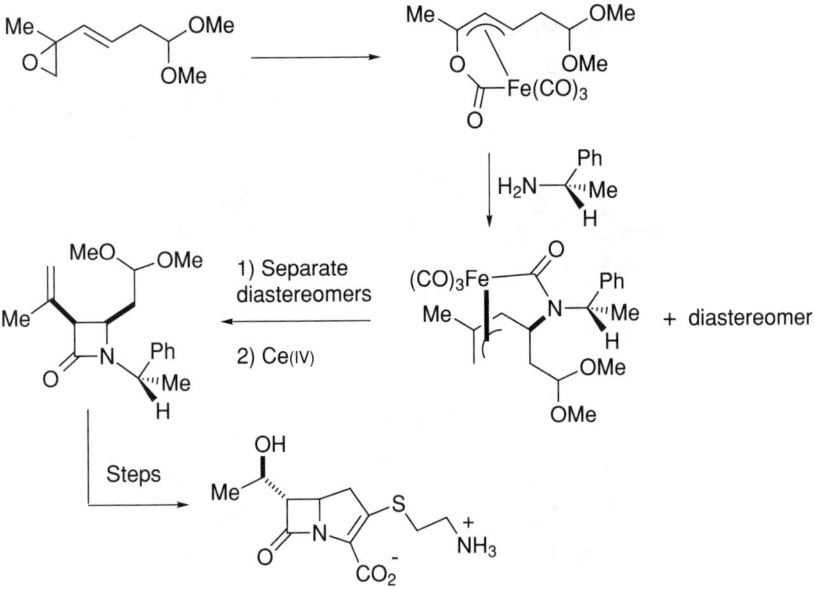

3.66

SCHEME 3.4. Synthesis of (+)-thienamycin.

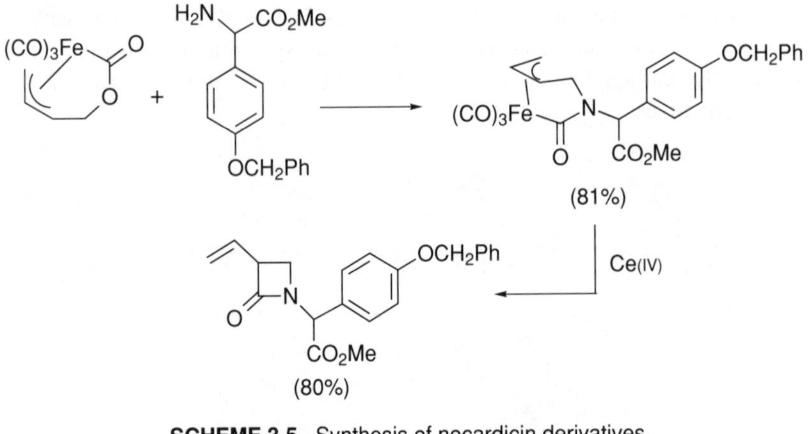

(81%)

(80%)

SCHEME 3.5. Synthesis of nocardicin derivatives.

The conversion of ferrelactone complexes into δ-lactones, by heating under CO pressure, has also been exploited for natural products synthesis. One application is in the synthesis of malyngolide (**3.68**, Scheme 3.6)[33], an antibiotic isolated from the blue–green alga *Lyngbya majuscula*. In this synthesis, both ferrelactone diastereomers could be converted into racemic malyngolide, since epimerization alpha to the lactone carbonyl of the undesired product **3.67** is possible.

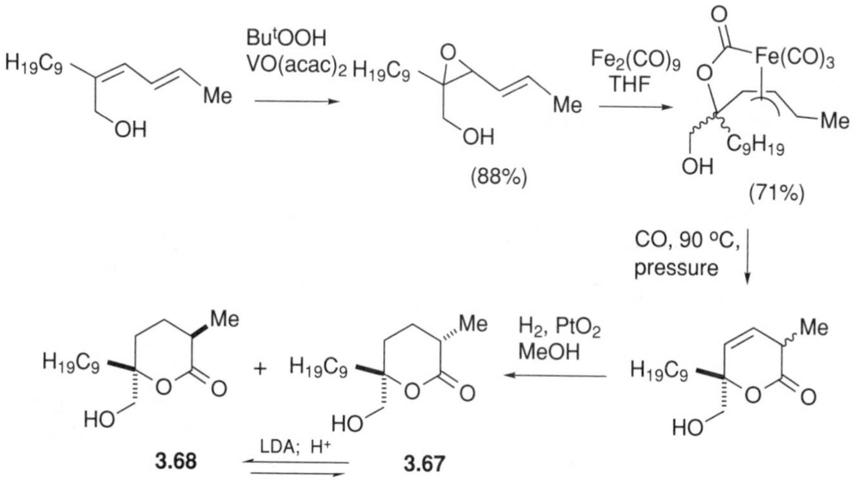

Bu^{t}OOH VO(acac)$_2$; (88%); $Fe_2(CO)_9$ THF; (71%); CO, 90 °C, pressure; H_2, PtO_2 MeOH; LDA; H$^+$; **3.68**; **3.67**

SCHEME 3.6. Synthesis of (rac)-malyngolide.

This approach to the synthesis of δ-lactones was also used to assemble one of the building blocks for Ley's total synthesis of Avermectin B1α[34], a member of a group of compounds used in the control of parasitic diseases. Since this forms a small part of the overall effort, the entire synthesis will not be described here, but attention will be confined to the organoiron chemistry that was used to construct the C(21)–C(25) ring structure (see Avermectin structure **3.69**). The intermediate required for this was the sulphone tetrahydropyran **3.70** which was synthesized as shown in Scheme 3.7. The major problem associated with the synthesis was the formation of a mixture of stereoisomeric lactones **3.73, 3.74** and **3.75**. Different ratios of these were obtained from ferrelactones **3.71** (40:7:51) and **3.72** (65:24:10) and the desired compound **3.73** could be obtained in 52% combined yield after removal of the by products by chromatography. Hydrogenation of this compound gave **3.76**, which was readily converted into the sulphone intermediate **3.70**.

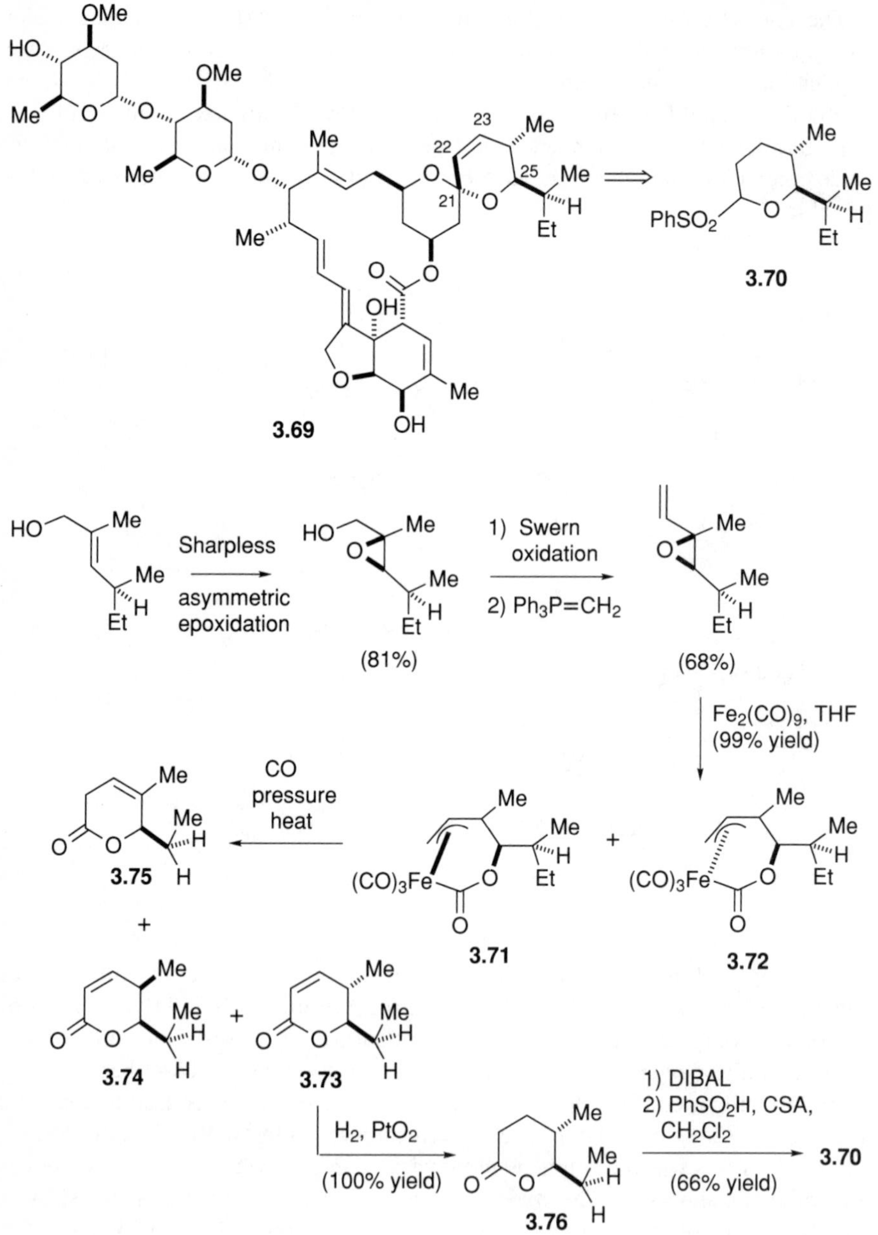
3.69
3.70
Sharpless asymmetric epoxidation
(81%)
1) Swern oxidation
2) Ph3P=CH2
(68%)
Fe2(CO)9, THF (99% yield)
3.71
3.72
CO pressure heat
3.75
3.74
3.73
H2, PtO2
(100% yield)
3.76
1) DIBAL
2) PhSO2H, CSA, CH2Cl2
(66% yield)
3.70

SCHEME 3.7

3.3 η^1-ACYLIRON COMPLEXES

Alkylation of $[CpFe(CO)_2]^-$ with methyl iodide or reaction of $CpFe(CO)_2I$ with MeMgI gives $Cp(CO)_2FeCH_3$ (**3.77**) which, on treatment with triphenylphosphine, undergoes migratory insertion and phosphine ligand incorporation to generate the chiral-at-metal acyliron complex **3.78**. The acyl group in this complex undergoes many of the reactions associated with ketone enolates and, as we shall see, the chiral metal center leads to excellent levels of asymmetric induction during alkylation and other reactions, making them very attractive intermediates for organic synthesis. They have been studied extensively by the research groups of Davies and Liebeskind. Davies has utilized optically pure complexes, obtained by the resolution of the racemic compounds, in a number of elegant synthetic applications. Although full details of the resolution procedure have not been disclosed in the chemical literature, the optically pure compounds are now commercially available (Oxford Asymmetry), but they are rather expensive.

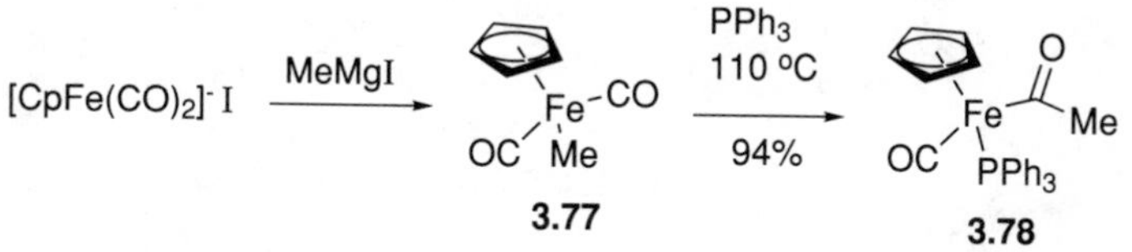

$[CpFe(CO)_2]^-$ I —MeMgI→ **3.77** —PPh₃, 110 °C, 94%→ **3.78** (3.41)

Figure 3.1(a) shows the structure of (R)-(−)-**3.78**, which has pseudooctahedral geometry in which the triphenylphosphine ligand is pushed close to the acyl group[39–40]. This results in a very effective shielding of one face of an enolate derived from the complex (Fig. 3.1b), so that electrophiles preferentially approach the top face as indicated.

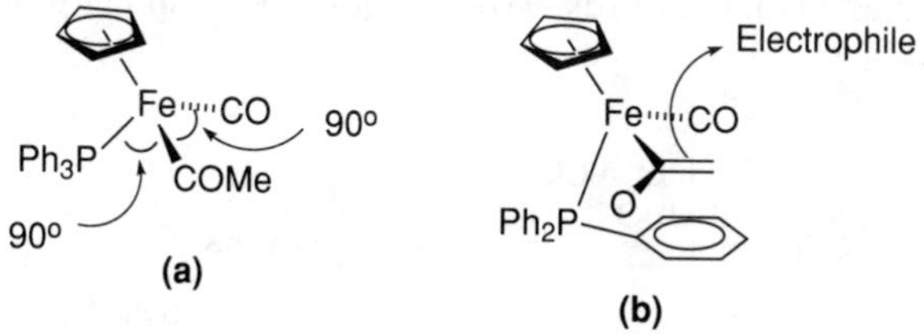

FIG. 3.1. Structure of (R)-(−)-**3.78** and derived enolate.

Early work has described diastereoselective reactions of racemic complexes[41–45]; attention here will be confined mostly to the more recent work using optically pure complexes. Treatment of **3.78** with Bu^{n}Li, followed by methyl iodide affords the propionyl complex **3.79** in 99% yield. Reaction of this compound with base, followed by enolate alkylation using t-butyl bromomethyl thioether gives (*R,S*)-**3.80** as a single diastereomer in 71% yield. The product is demetallated using bromine followed by L-proline t-butyl ester to give the amide **3.81** in 83%

yield, deprotection of which affords the angiotensin converting enzyme inhibitor (−)-captopril (**3.82**) in quantitative yield (Scheme 3.8)[46].

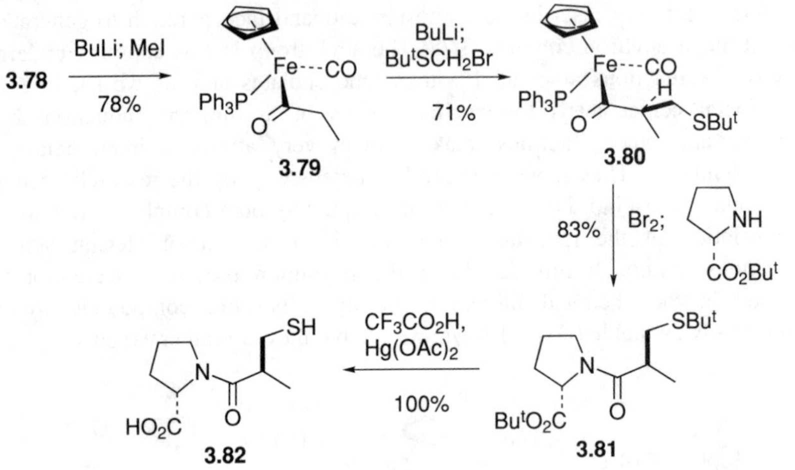

3.78 → (BuLi; MeI, 78%) → **3.79** → (BuLi; Bu^tSCH_2Br, 71%) → **3.80** → (Br_2; proline Bu^t ester, 83%) → **3.81** → (CF_3CO_2H, $Hg(OAc)_2$, 100%) → **3.82**

SCHEME 3.8

One of the attributes of enolate alkylations on **3.78**, that is not found with other asymmetric enolates, is its ability to undergo diastereoselective reaction with aldehydes, without the need for methyl substitution on the α-carbon. Stereocontrol better than 100:1 has been obtained during such reactions (equation 3.42)[47,48]. Enolate generation from **3.83** (using two equivalents of BuLi), followed by methylation, affords *erythro* aldol-type coupling products of general structure **3.84**.

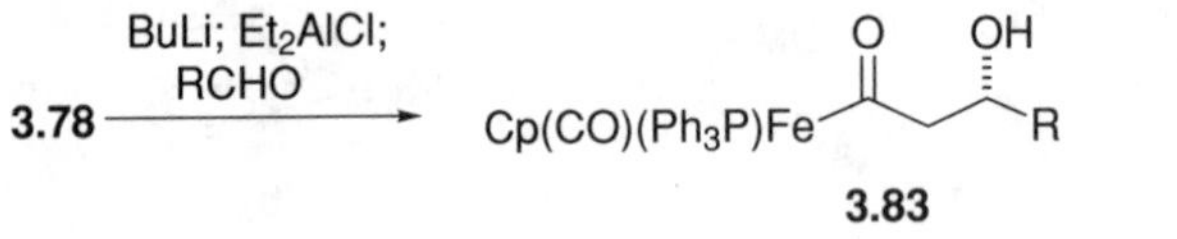

3.78 → (BuLi; Et_2AlCl; RCHO) → Cp(CO)(Ph_3P)Fe–CO–CH_2–CH(OH)R **3.83** (3.42)

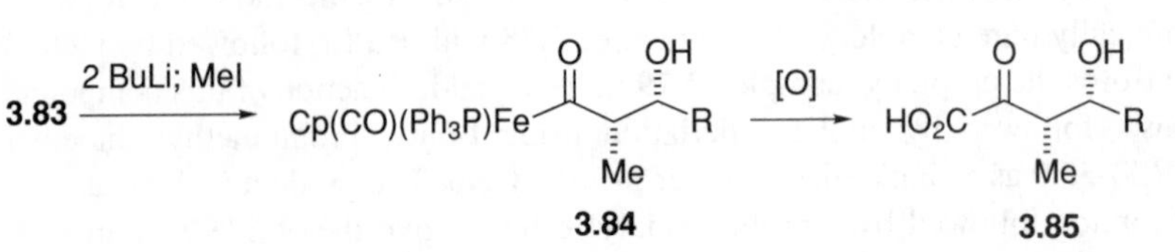

3.83 → (2 BuLi; MeI) → Cp(CO)(Ph_3P)Fe–CO–CH(Me)–CH(OH)R **3.84** → ([O]) → HO_2C–CH(Me)–CH(OH)R **3.85** (3.43)

The nature of the countercation used with enolates from propionyl-Fe derivatives (**3.79**) markedly affects the stereochemical outcome of the aldol reactions[49]. For example, reaction of the lithium enolate **3.86** with aldehydes is essentially stereorandom at −100°C, giving all four possible diastereomers as shown in equation 3.44. If the lithium enolate is treated with cuprous cyanide (1 equiv., −40°C, 2 h) the copper enolate is generated. Reaction of this with aldehydes at −78°C gives predominantly **3.88** together with small amounts of **3.87**. In contrast, conversion of **3.86** into the diethylaluminum enolate (3 equiv. Et_2AlCl, −40°C, 1.5 h) prior to reaction with aldehydes at −100°C allows the formation of predominantly **3.87** together with small amounts of **3.89**.

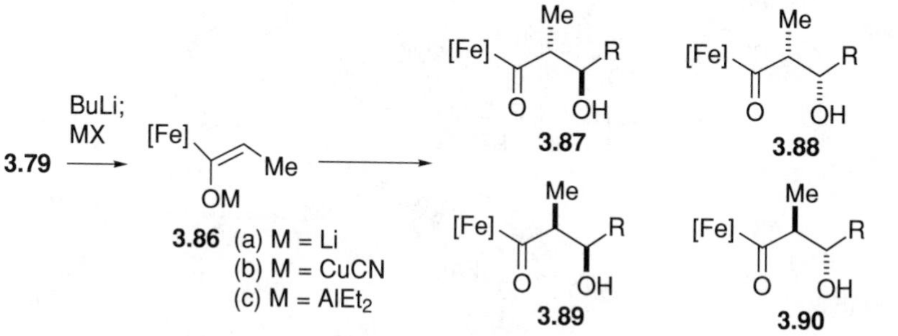

BuLi; MX
3.79 → [Fe] (OM) =\ Me → **3.87**, **3.88**, **3.89**, **3.90** (3.44)

3.86 (a) M = Li
(b) M = CuCN
(c) M = $AlEt_2$

These results have been explained on the basis of *syn* and *anti* conformations of the respective enolates (Fig. 3.2). The lithium enolate reacts non-stereoselectively *via* both *anti* and *syn* enolates, whereas the aluminum enolate reacts stereoselectively on both conformations; addition to the *anti* enolate is preferred and this leads to the formation of **3.87**. The copper enolate also reacts via the *anti* enolate, but the *erythro* transition state is favored, leading to **3.88**. The steric bulk of the aldehyde R group affects the ratio of **3.88**:**3.87** in the latter reaction; as R is made larger the ratio increases in favor of **3.88**. In general the aluminum enolates show complete stereocontrol from the iron moiety over the β-center, with moderate to excellent stereocontrol over the α-center. The copper enolates, on the other hand, appear to give complete stereocontrol over the α-center, while β stereoselectivity increases with the steric demand of the aldehyde R group.

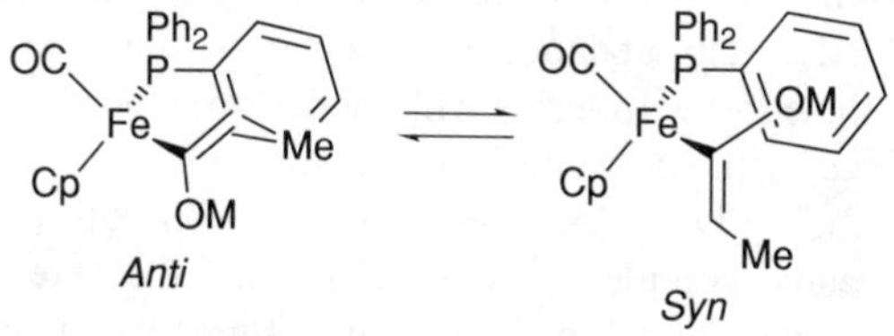

FIG. 3.2. *Anti* and *syn* conformations of enolates **3.86**.

Another interesting application of the chiral acyliron complexes is in the preparation of chiral sulfoxides. Reaction of the lithium enolate **3.86a** (optically pure, *R* configuration) with diphenyldisulfide furnishes **3.91** as a 16:1 mixture of diastereomers favoring **3.91a** (Scheme 3.9). A single crystallization gives diastereomerically pure material which is oxidized to **3.92**, also obtained as a single diastereomer[50,51]. Reaction of **3.92** with dibutylcopperlithium gives optically pure sulfoxide **3.93** together with complex **3.79** which is therefore recycleable (unreacted complex **3.92** is also obtained as a result of competing enolization). Oxidative demetallation of **3.92** in the presence of benzylamine gives the β-sulfinylamide **3.94**.

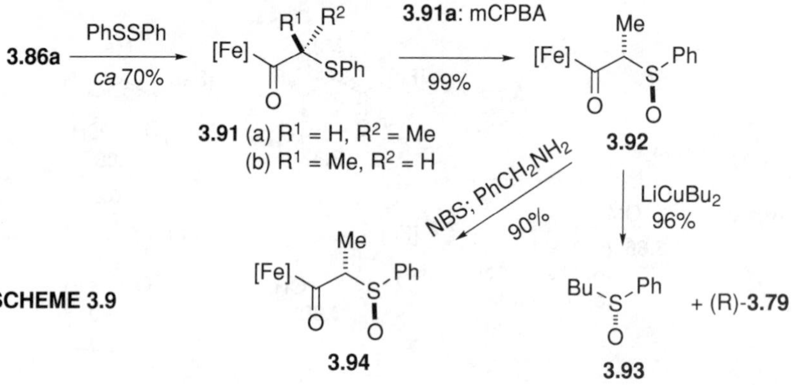

SCHEME 3.9

Extension of the aldol reaction shown in equation 3.42, using an imine instead of aldehyde, leads to a method for the construction of β-lactam derivatives (equation 3.46). During the decomplexation the activated acyliron reacts intramolecularly with the pendant amine[52].

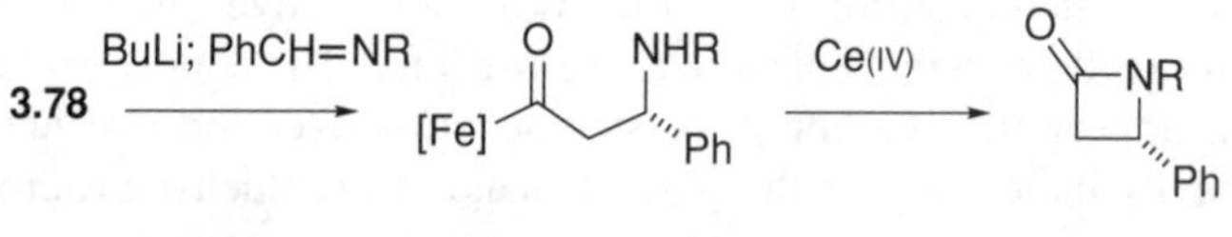

(3.46)

An interesting example of remote stereocontrol is shown in Scheme 3.10[53,54], which involves selective deprotonation of a succinoyliron complex alpha to the ester. Alkylation of the enolate proceeds with modest to good stereoselectivity, best results being obtained with more sterically demanding alkyl halides (e.g., MeI gives 3:1 while Bu^iI gives 15:1). Decomplexation of the recrystallized, diastereomerically pure complex, using NBS in the presence of (+)-α-methylbenzylamine, proceeds with maintenance of stereochemical integrity to give **3.97** in high yield.

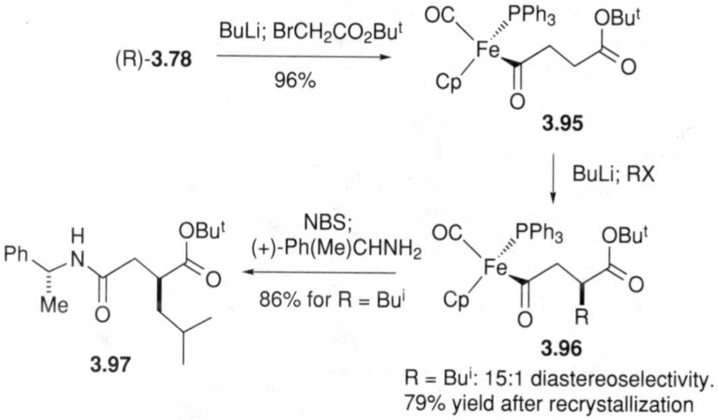

SCHEME 3.10

Similarly, "remote" stereocontrol has been observed during conjugate additions to α,β-unsaturated acyliron complexes[55,56]. *In situ* enolate alkylation allows the construction of vicinal stereocenters with outstanding diastereoselectivity (equation 3.47).

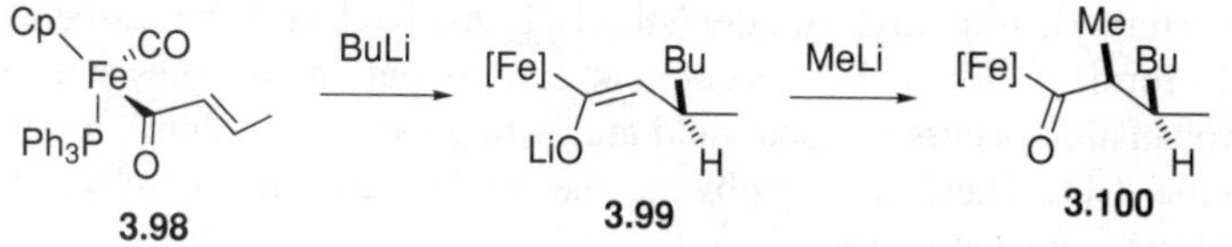

(3.47)

3.4 CARBENE COMPLEXES OF IRON

Iron-carbene complexes have received much less attention than has been devoted to chromium-carbene derivatives, but they offer a number of interesting and potentially useful reactions. Pettit[57] observed that reaction of sodium dicarbonyl-η^5-cyclopentadienylferrate (NaFp, see equation 3.1) with chloromethyl methyl ether gave the σ-methoxymethyliron complex **3.101**. When this complex was treated with acid in the presence of an alkene, cyclopropanes were formed (equations 3.48–3.50), implying the intermediacy of a cationic carbene complex **3.102**. The related phenylcarbenium complexes **3.103** and **3.104** were later isolated and characterized using ^{1}H NMR spectroscopy[58].

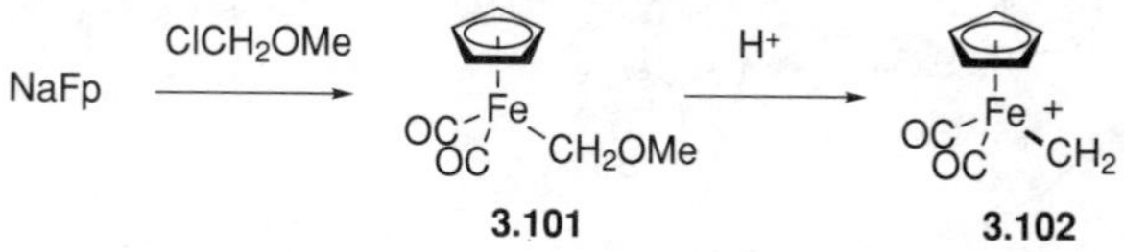

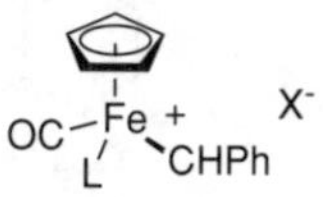

3.103 L = CO
3.104 L = PPh_3

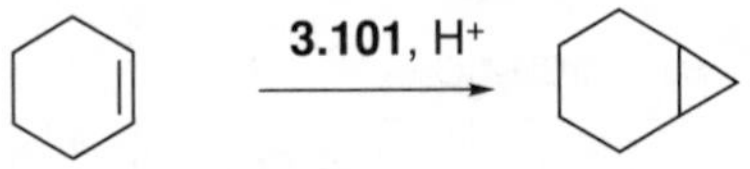

3.101, H^+ (3.48)

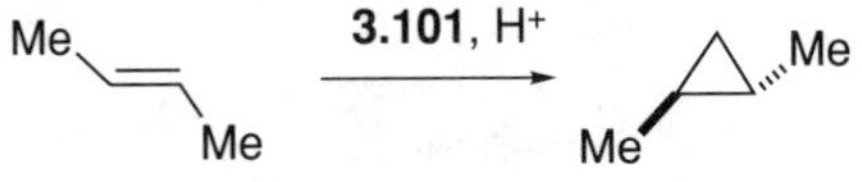

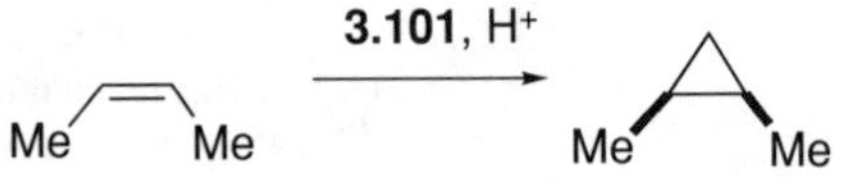

3.101, H^+ (3.49)

Me Me Me Me

3.101, H^+ (3.50)

Me Me Me Me

The reactions shown in equations 3.48–3.50 are not particularly general, but methods have been developed for improving the cyclopropanation[59,60]. The thioether complex **3.105** can be methylated to give the sulfonium derivative **3.106**, which is a stable isolable compound (Scheme 3.11). When **3.106** is heated in dioxane (reflux), dimethyl sulfide (stink!) is evolved and the carbene complex **3.102** is formed. When this reaction is carried out in the presence of olefins, cyclopropanation occurs in good yield and with good stereocontrol, as summarized in Scheme 3.11. These tactics obviate the use of acid and thereby open up the method for more general use.

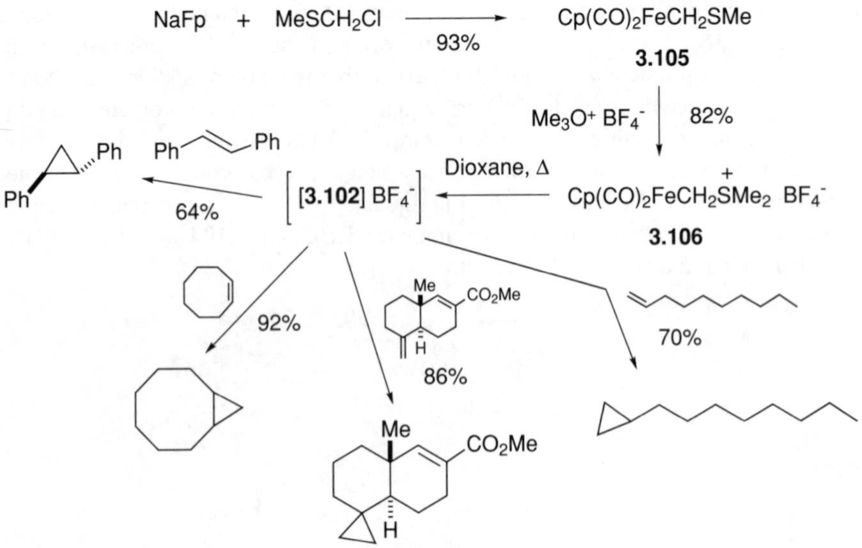

NaFp + $MeSCH_2Cl$ → (93%) $Cp(CO)_2FeCH_2SMe$ **3.105**

$Me_3O^+ BF_4^-$ 82%

$Cp(CO)_2FeCH_2\overset{+}{S}Me_2\ BF_4^-$ **3.106**

Dioxane, Δ

[[**3.102**] BF_4^-]

Ph Ph Ph Ph 64%

92%

Me CO_2Me H 86%

70%

Me CO_2Me H

SCHEME 3.11

Further work by Helquist resulted in the development of methods for the preparation of a variety of substituted iron-carbene complexes[61]. Treatment of the phenylthiomethyl-Fe(CO)$_2$Cp complex **3.107** with trityl hexafluorophosphate gives the stable (phenylthio)carbene complex **3.108** in 80% yield after recrystallization. Other alkylthiocarbene complexes (**3.109**) can be prepared as shown in equation 3.52. These thiocarbene complexes react with nucleophiles to give more highly substituted alkyl-Fe(CO)$_2$Cp derivatives (equations 3.53–3.55), which are potential precursors to a variety of functionalized carbene complexes.

$$\underset{\textbf{3.107}}{Cp(CO)_2FeCH_2SPh} \xrightarrow[80\%]{Ph_3CPF_6,\ CH_2Cl_2,\ rt} \underset{\textbf{3.108}}{[Cp(CO)_2Fe{=}CHSPh]^+\ PF_6^-} \quad (3.51)$$

$$Cp(CO)_2FeCOCH_3 \xrightarrow{Tf_2O,\ -78^oC} \left[\begin{matrix} Cp(CO)_2\overset{+}{Fe}{=}C{=}CH_2 \\ CF_3SO_3^- \end{matrix}\right] \xrightarrow[-78^oC]{R'SH,\ ca\ 70\%} [Cp(CO)_2Fe{=}CMeSR']^+\ X^- \quad (3.52)$$

3.109 (a) R' = Me
(b) R' = Ph
$X^- = CF_3SO_3^-$ or PF_6^-

$$\underset{\textbf{3.108}}{[Cp(CO)_2Fe{=}CHSPh]^+\ PF_6^-} \xrightarrow{R\text{-}M,\ THF,\ -78^oC} \underset{\textbf{3.110}}{Cp(CO)_2FeCH(R)SPh} \quad (3.53)$$

(a) R-M = MeLi (93%)
(b) R-M = MeMgBr (73%)
(c) R-M = n-C_5H_{11}MgBr (70%)
(d) R-M = PhMgBr (70%)

$$\text{(1-lithiooxycyclohexene)} \xrightarrow[91\%]{\textbf{3.108},\ THF,\ -78^oC} \text{2-[CH(SPh)Fe(CO)}_2\text{Cp]cyclohexanone} \quad (3.54)$$

3.111 (3:2 diastereomers)

$$\textbf{3.109 a, b} \xrightarrow{LiCuR_2} \underset{\textbf{3.112}}{Cp(CO)_2FeC(SR')(R)Me} \quad (3.55)$$

(a) R = R' = Me (63%)
(b) R = Bu^n, R' = Me (51%)
(c) R = Me, R' = Ph (60%)

Br; Me; 1) Mg; 2) **3.108**; SPh; $Fe(CO)_2Cp$; Me; Me_3OBF_4; H; Me; 60% overall (3.56)

O; H; $Fe(CO)_2Cp$; SPh; H; Me_3OBF_4; 51%; O; H; H; H; H; +; O; H; H; (3:2) (3.57)

O; H; $Fe(CO)_2Cp$; SPh; H; Me_3OBF_4, CH_2Cl_2, 0 – 25°C; 90%; O; H; H (3.58)

O; H; $Fe(CO)_2Cp$; SPh; H; Me_3OBF_4; O; H; Et; H; +; O; H; Et; H (3.59)

The above methodology allows access to carbene complexes that can be used in intramolecular reactions with alkenes[62,63]. These reactions, however, do not always lead to intramolecular cyclopropanation but to cationic annulation or carbene C–H insertions, examples of which are shown in equations 3.56–3.59. The C–H insertion reactions (equations 3.57–3.59) are potentially very useful, since they occur with unactivated C–H bonds and with good regiocontrol, forming five-membered rings. One illustration of the reaction's potential utility is outlined in Scheme 3.12, leading to a total synthesis of the fungal metabolite sterpurine[64].

The use of related vinylideneiron complexes in the synthesis of β-lactam derivatives has been investigated[65–67]. Some examples of this chemistry, having obvious implications for the preparation of penicillin analogs, are given in Scheme 3.13, which shows the stepwise nature of the bond-forming process indicating a possible mechanism for the overall conversion. The conversion of **3.114** into **3.115** proceeds with good to excellent diastereoselectivity (6:1 to 15:1), the major isomer having the stereochemistry shown for **3.116**.

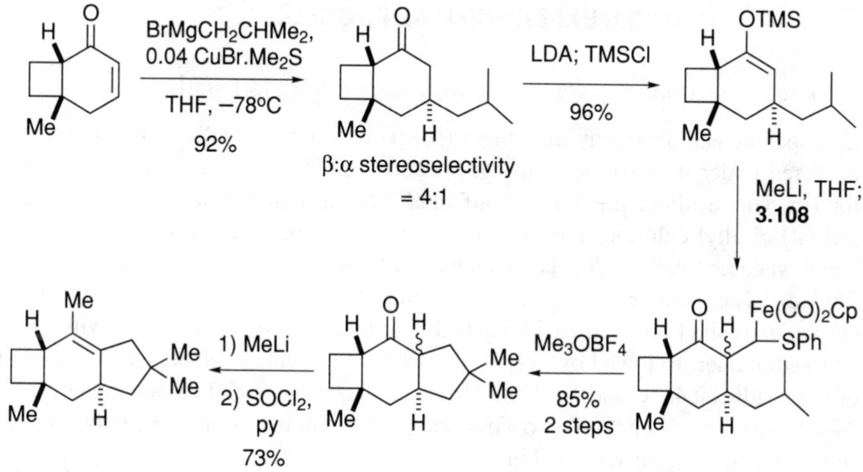

H O Me

BrMgCH$_2$CHMe$_2$, 0.04 CuBr.Me$_2$S

THF, −78°C

92%

H O Me H

β:α stereoselectivity = 4:1

LDA; TMSCl

96%

OTMS H Me H

MeLi, THF; **3.108**

O Fe(CO)$_2$Cp H H SPh Me H

Me$_3$OBF$_4$

85%

2 steps

H O H Me Me Me H

1) MeLi

2) SOCl$_2$, py

73%

Me H Me Me Me H

SCHEME 3.12. Total synthesis of (±)-sterpurine using iron-carbene intermediates.

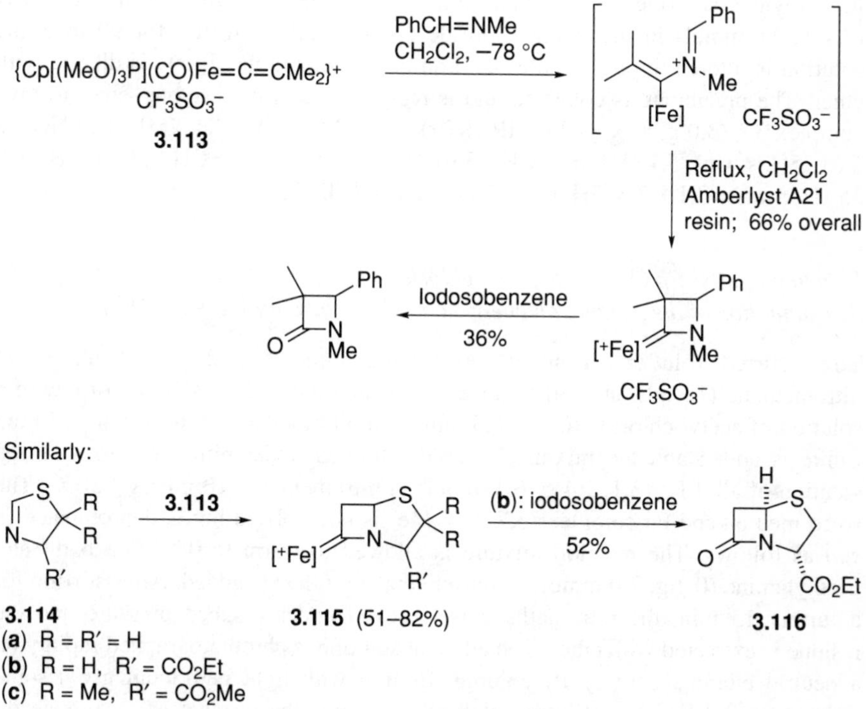

{Cp[(MeO)$_3$P](CO)Fe=C=CMe$_2$}$^+$

CF$_3$SO$_3^-$

3.113

PhCH=NMe

CH$_2$Cl$_2$, −78 °C

Ph $^+$N Me [Fe] CF$_3$SO$_3^-$

Reflux, CH$_2$Cl$_2$

Amberlyst A21

resin; 66% overall

Ph N Me O

Iodosobenzene

36%

Ph [$^+$Fe] N Me

CF$_3$SO$_3^-$

Similarly:

S R R N R′

3.113

S R R [$^+$Fe] N R′

(**b**): iodosobenzene

52%

H S N O CO$_2$Et

3.114
(**a**) R = R′ = H
(**b**) R = H, R′ = CO$_2$Et
(**c**) R = Me, R′ = CO$_2$Me

3.115 (51–82%)

3.116

SCHEME 3.13. Uses of vinylideneiron complexes in β-lactam synthesis.

3.5 EXPERIMENTAL PROCEDURES

Preparation of σ-allyl(cyclopentadienyl)dicarbonyliron (**3.1**)[68]

Cyclopentadienyldicarbonyliron dimer (see p. 32, 7.1 g) in THF solution (*ca* 50 ml) is stirred under nitrogen at room temperature with 3% sodium–mercury amalgam for 1 h. The resulting solution of $Na[CpFe(CO)_2]$ is added slowly to a large excess (30 ml) of allyl chloride, and the mixture is stirred for 1 h. Volatiles are removed under vacuum, the residue is extracted with light petroleum, and the slurry is filtered. The filtrate is evaporated and the resulting red–brown oil is chromatographed on an acid-washed alumina column. Elution with light petroleum/ether, followed by evaporation of the light yellow eluate, and distillation of the residue at 45°C and 10^{-3} mm Hg gives the complex (**3.1**) as an amber oil (*ca* 34% yield), dec *ca* 65°C. The σ-allyl complex is soluble in most common organic solvents, and is rapidy oxidized in air.

Reaction of allyl-Fp (**3.1**) *with trimethyloxonium tetrafluoroborate (equation 3.5)*[69]

To a stirred suspension of trimethyloxonium tetrafluoroborate (1.85 g, 12.5 mmol) in methylene chloride (20 ml) under nitrogen is added a solution of allyl-Fp (**3.1**) (2.4 g, 11 mmol) in methylene chloride (10 ml). After stirring for 90 min, the solution is filtered through a short column of Celite, and the filtrate is diluted with ether. The precipitate is collected and is recrystallized from acetone–ether to give complex **3.6** (3.0 g, 85% yield). IR (KBr) 2080, 2040 cm^{-1}. ^1H NMR (CD_3NO_2) δ 5.61 (5H, s, Cp), 5.1 (1H, m, =CH), 3.91 (1H, d, *J* 8 Hz, *cis* $=CH_2$), 3.46 (1H, d, *J* 15 Hz, *trans* $=CH_2$), 2.4 (2H, m), 1.16 (3 H, t, *J* 7 Hz, Me).

Reaction of allyl-Fp (**3.1**) *with acetyl chloride (equation 3.6, R = Me) and deprotonation of the product alkene-Fp complex (equation 3.2 + 3.3)*[69]

To a stirred solution of silver hexafluoroantimonate (2.2 g, 6.4 mmol) in nitromethane (15 ml), at −30°C under nitrogen atmosphere, is added dropwise a solution of acetyl chloride (0.5 g, 6.5 mmol) in nitromethane (5 ml). After 15 min, Celite is added and the mixture is rapidly filtered under nitrogen into a stirred solution of allyl-Fp (**3.1**, 1.4 g, 6.4 mmol) in nitromethane (10 ml) at −30°C. The so-formed alkene-Fp complex (**3.7**, R = Me) is not isolated but is deprotonated *in situ* as follows. The reaction mixture is allowed to warm to 0 °C (ice bath) and triethylamine (0.7 g, 7.0 mmol) in nitromethane (10 ml) is added. After stirring for a further 15 min, the nitromethane is removed under reduced pressure and the residue is extracted with ether. The ethereal solution is chromatographed rapidly on a neutral alumina activity III column. Elution with light petroleum gives some unreacted allyl-Fp (**3.1**). Elution with ether affords the product as an orange oil (equation 3.2, X = COMe). An analytical sample can be prepared by recrystallization from carbon disulfide at −45°C to give orange crystals, m.p. 34.5–36°C. IR (CH_2Cl_2) 2010, 1950, 1625 cm^{-1}. ^1H NMR (CS_2) δ 7.0 (1H, dt, *J* 15, 9 Hz), 5.7 (1H, d, *J* 15 Hz), 4.7 (5H, s, Cp), 2.05 (2H, d, *J* 9 Hz, CH_2), 2.0 (3H, s, Me).

[3 + 2] Cycloaddition reaction of allyl-Fp with tetracyanoethylene

A solution of tetracyanoethylene (97 mg, 0.75 mmol) in degassed benzene (8 ml) is added dropwise to a solution of the acetylallyl complex from above (190 mg, 0.75 mmol) in benzene (3 ml). After standing for 2 h, a yellow crystalline solid precipitates out. Petroleum ether (12 ml) is added in order to complete the precipitation and the product is collected by filtration and washed with several portions of petroleum ether (200 mg, 70% yield). An analytical sample may be obtained by slow evaporation of a methylene chloride–hexane solution using a nitrogen stream, to give yellow needles, m.p. 150–151°C, dec. IR (KBr) 2375, 2020, 1950, 1720 cm^{-1}.

Preparation of π-allylirontetracarbonyl bromide[11]

A suspension of nonacarbonyldiiron (36.5 g, 0.1 mol) in hexane (50 ml) is treated with allyl bromide (12.0 g, 0.1 mol) and warmed with stirring at 40°C. After 90 min, a dark-red solution results, that is filtered away from the iron(II) bromide formed by partial decomposition. After removal of the pentacarbonyliron *(CAUTION! See p. 90 for handling of pentacarbonyliron)* and of the solvent at approx. 20–30°C/12 Torr and standing for a while, the product crystallizes as yellow–brown prisms (4.0 g). By further concentration of the mother liquor another 6.1 g are obtained, and the combined fractions are recrystallized from petroleum ether (b.p. 45–60°C). The yield is 9.9 g (38% of the theoretical yield).

Conversion of vinyloxiranes to ferrelactones (equation 3.38)[70]

3,4-Epoxycyclohex-l-ene is prepared as follows: Cyclohexa-1,3-diene (15 g, 0.18 mol) is added to a stirred suspension of sodium carbonate (50 g) in methylene chloride (200 ml) at 0°C. To the stirred suspension is added dropwise peracetic acid (38.8 g, 38%), containing a catalytic amount of sodium acetate, over a period of 1 h. After completion, the mixture is allowed to warm to room temperature, filtered, and the precipitate is washed with methylene chloride. Removal of the solvent from the combined filtrate and washings gives an oil which is distilled under reduced pressure to afford the vinyloxiran (10.3 g, 59% yield), b.p. 77–78°C at 93 mm Hg. A solution of the vinyloxiran (1 g, 10.4 mmol) and pentacarbonyliron (10 g, 51.0 mmol., *CAUTION! See p. 90*) in de-gassed benzene (400 ml) is irradiated at room temperature under an argon atmosphere with two external 450-W Hanovia medium-pressure mercury lamps through Chance OX 1 filters. The reaction is followed by IR until the formation of the complex is maximized (2–6 h). Solvent and excess pentacarbonyliron (*CAUTION!*) are removed under vacuum, taking care not to heat the mixture above 10°C. The residue is stirred with ether and filtered through a pad of Celite. The filtrate and washings are reduced in volume, triturated with petroleum ether and the product is removed by filtration to give the ferrelactone complex (0.86 g, 32% yield) m.p. 74°C (dec) (Aumann *et al.*[28,29] report the m.p. of this complex as 80°C) IR: 2100, 1990, 1650 cm^{-1} (diagnostic peaks)

Conversion of ferrelactone complexes into β-lactones (equation 38)[70]

The ferrelactone complex (290 mg, 1.09 mmol, see above) is added in one portion to a stirred slurry of cerium (IV) ammonium nitrate (3.61 g, 6.6 mmol) in ethanol (15 ml) at −5°C. After complete oxidation (2–6 h), as indicated by TLC, the solvent is removed under reduced pressure. The residue is mixed with the minimum amount of water to dissolve inorganic materials and the mixture is extracted with ether (3 × 5 ml). The combined extracts are dried (Na_2SO_4) and the solvent is removed under reduced pressure to afford 7-oxabicyclo[4.2.0]oct-2-en-8-one as an oil which can be purified by chromatography (116. 8 mg, 86% yield). IR: 1815 cm^{-1}.

Preparation of racemic acetyl(carbonyl)(cyclopentadienyl)(triphenylphosphine)iron **(3.78)**

The procedure for the reaction of $CpFe(CO)_2I$ with MeMgI is taken from ref. [71]; the procedure for conversion of $CpFe(CO)_2Me$ into **3.77** is taken from ref. [72]. $CpFe(CO)_2I$ in dilute ether solution (0.02–0.1 M) is treated with an ether solution of methylmagnesium iodide. A stoichiometric amount of the Grignard reagent is used and is added to the halide solution with brisk stirring, under nitrogen atmosphere (using syringe for small scale). After the addition is complete, the mixture is stirred rapidly for 2–3 h at 25°C and the solvent is then removed under vacuum. The residue is subjected to vacuum sublimation onto an ice cooled probe, giving complex **3.77** in *ca* 50% yield as a caramel colored solid, m.p. 78–82°C, IR 2120 (w), 2010(vs), 1955(vs), 1920(w) cm^{-1}. This complex (1.0 g, 5 mmol) and triphenylphosphine (1.31 g, 5 mmol) are dissolved in freshly distilled THF and refluxed under dry nitrogen. Progress of the reaction can be monitored by IR spectroscopy (product Fe*CO* at 1920, Me*CO* at 1590 cm^{-1}). When no more $CpFe(CO)_2Me$ can be detected (*ca* 48 h), the mixture is filtered to remove trace amounts of an insoluble brown solid. The solvent is removed under reduced pressure and the residue is dissolved in *ca* 10 ml of pentane. Chromatography on an alumina column (5 × 20 cm) gives a single orange band that is eluted with pentane. Removal of solvent under reduced pressure gives the racemic complex **3.78**, m.p. 145°C (96–99% yield).

Alkylation of enolate from complex **3.78**[41,42]

Diisopropylamine (190 µl, 1.36 mmol) is added to a solution of n-butyllithium (825 µl of 1.6 M hexane solution, 1.32 mmol) in THF (8 ml) maintained at 0°C under nitrogen. After stirring the mixture at 0°C for 20 min, the flask is cooled in a solid CO_2/acetonitrile bath (*ca* −42°C). To this LDA solution is added, dropwise by syringe, a solution of the iron acyl **3.78** (500 mg, 1.10 mmol) in THF (4 ml). The resulting deep red colored solution is stirred at −42°C for 40 min, at which time the enolate is quenched with an appropriate alkyl halide. For example, addition of benzyl bromide (170 µl, 1.43 mmol) causes an immediate color change from red to dark orange. The reaction mixture is allowed to warm to room temperature, and after 30 min aqueous ammonium chloride (5%, 30 ml) is added, followed by methylene chloride (30 ml) and 1.2 N HCl (20 ml). The organic layer is separated, the aqueous layer is extracted with methylene chloride (2 × 30 ml) and the combined organic phase is dried (Na_2SO_4) and evaporated under vacuum. The resulting orange oil is purified by chromatography on alumina (2 × 15 cm, CH_2Cl_2) to give carbonyl(cyclopentadienyl)(triphenylphosphine)(3-phenylpropionyl)iron (542 mg, 91% yield) as a deep orange solid, m.p. 128–131°C (from methylene chloride/petroleum ether). IR ($CHCl_3$) 1920, 1600 cm^{-1}. The propionyl complex **3.79** is prepared in 78% yield in an analogous fashion, using methyl iodide in place of benzyl bromide.

Cleavage of the metal-acyl bond of acyl-Fe(CO)(PPh$_3$)Cp complexes[41,42]

The phenylpropionyl complex from above (389 mg, 0.715 mmol) is dissolved in methylene chloride (5 ml) and diluted with an equal volume of ethanol. The stirred solution is cooled in a solid CO_2/acetonitrile bath (*ca* −42°C) and a solution of *N*-bromosuccinimide (142 mg, 0.793 mmol) in methylene chloride (5 ml) is added dropwise. The color of the solution slowly changes from orange to deep green. After the addition the solution is allowed to warm to room temperature and after 45 min is transferred to a separatory funnel with the aid of methylene chloride, and then washed with 1 N NaOH. The aqueous wash is extracted with methylene chloride and the combined organic layers are dried (Na_2SO_4) and then filtered. Removal of the volatiles on a rotary evaporator at room temperature leaves a green semi-solid residue that is triturated several times with ether. The combined ether extracts are condensed to an oil which is filtered through a short alumina column with petroleum ether to give ethyl dihydrocinnamate (114 mg, 89% yield).

Preparation of cyclopentadienyl(dicarbonyl)[(phenylthio)carbenium]iron hexafluorophosphate (**3.108**)[61]

The following description for the preparation of the precursor **3.107** is adapted from the procedure for making the methylthiomethyl analog **3.105**, described in detail in ref. [73]. A solution of 100 mmol of $NaFe(CO)_2Cp$ in 250 ml of THF (from Na/Hg reduction of the dimer as described on pp. 32–33) is treated dropwise with a solution of 15.9 g (100 mmol) of chloromethylphenylsulfide in THF (100 ml). An exothermic reaction occurs with very little color change. After stirring overnight at room temperature the solvent is removed at 25°C (25 mm). The residue is extracted with methylene chloride (3 × 100 ml) and the extracts are filtered first by suction through *ca* 20 g of chromatography-grade alumina and then by gravity. The solvent is removed from the filtrate at 25°C (25 mm) and the residue is extracted with pentane (3 × 50 ml). The solution is filtered and cooled to −78°C to effect crystallization of the complex (**3.107**), which is removed by filtration. *In the glove box* complex **3.107** (1.48 g, 5.0 mmol) is placed in a 100 ml flask equipped with a magnetic stirring bar. Methylene chloride (20 ml) is added, followed by trityl hexafluorophosphate (1.7 g, 5.0 mmol, p. 155). The orange solution immediately turns dark green and the mixture is magnetically stirred for 5 min. Anhydrous ether (40 ml) is added slowly to precipitate complex **3.108**, which is collected by filtration through a medium frit followed by washing with ether (3 × 10 ml). The product is dried under vacuum to give **3.108** as golden yellow crystals, m.p. 126°C dec. (1.76 g, 80% yield). IR (CH_2Cl_2) 2073, 2034 cm^{-1}. ^1H NMR (CD_2Cl_2) δ 15.52 (s, $Fe^+{=}CH$), 7.62 (s, Ph), 5.64 (s, Cp).

References

1. M. Rosenblum, *Acc. Chem. Res.* **7**, 122 (1974).
2. R. Baker, C.M. Exon, V.B. Rao and R.W. Turner, *J. Chem. Soc. Perkin Trans.* **1**, 301 (1982).
3. R. Baker, V.B. Rao and E. Erdik, *J. Organomet. Chem.* **243**, 451 (1983).
4. T.S. Abram, R. Baker, C.M. Exon and V.B. Rao, *J. Chem. Soc., Perkin Trans 1*, 285 (1982).
5. R. Baker, R.B. Keen, M.D. Morris and R.W. Turner, *J. Chem. Soc. Chem. Commun.* 987 (1984).
6. W. P. Giering and M. Rosenblum, *J. Am. Chem. Soc.* **93**, 5299 (1971).
7. N. Genco, D. Marten, S. Raghu and M. Rosenblum, *J. Am. Chem. Soc.* **98**, 848 (1976).
8. J.C. Watkins and M. Rosenblum, *Tetrahedron Lett.* **25**, 2097 (1984).
9. J.C. Watkins and M. Rosenblum, *Tetrahedron Lett.* **26**, 3531 (1985).
10. B.M. Trost and T.R. Verhoeven, in *Comprehensive Organometallic Chemistry* (ed. G. Wilkinson, F.G.A. Stone and E.W. Abel), vol. 8, Chapter 57. Pergamon Press, Oxford, 1982.
11. R.A. Plowman and F.G.A. Stone, *Z. Naturforsch. Teil B.* **17**, 575 (1962).
12. H.D. Murdock and E. Weiss, *Helv. Chim. Acta* **45**, 1927 (1962).
13. T.H. Whitesides, R.W. Arhart and R.W. Slaven, *J. Am. Chem. Soc.* **95**, 5792 (1973).
14. G.F. Emerson and R. Pettit, *J. Am. Chem. Soc.* **84**, 4591 (1962).
15. D.H. Gibson and R.L. Vonnahme, *J. Am. Chem. Soc.* **94**, 5090 (1972).
16. D.H. Gibson and R.L. Vonnahme, *J. Organomet. Chem.* **70**, C33 (1974).
17. A.J. Pearson, *Aust. J. Chem.* **29**, 1841 (1976).
18. A.J. Pearson, *Tetrahedron Lett.* 3617 (1975).
19. J.L. Roustan, J.Y. Merour, and F. Houlihan, *Tetrahedron Lett.* 3721 (1979).
20. Y. Yu and B. Zhou, *J. Org. Chem.* **52**, 974 (1987).
21. Y. Yu and B. Zhou, *J. Org. Chem.* **53**, 4419 (1988).
22. J.L. Roustan, M. Abedini, and H.H Baer, *J. Organomet. Chem.* **376**, C20 (1989).
23. J.R. Green and M.K. Carroll, *Tetrahedron Lett.* **32**, 1141 (1991).
24. R.W. Bates, R. Fernandez-Moro and S.V. Ley, *Tetrahedron* **47**, 9929 (1991).
25. G.D. Annis and S.V. Ley, *J. Chem. Soc. Chem. Commun.* 581 (1977).
26. R.F. Heck and C.R. Boss, *J. Am. Chem. Soc.* **86**, 2580 (1964).

27. H.D. Murdoch, *Helv. Chim. Acta* **47**, 936 (1964).
28. R. Aumann, K. Frohlich and H. Ring, *Angew. Chem. Int. Ed. Engl.* **13**, 275 (1974).
29. R. Aumann, H. Ring, C. Kruger and R. Goddard, *Chem. Ber.* **112**, 3644 (1979).
30. K.-N. Chen, R.F.M. Moriarty, B.G. DeBoer, M.R. Churchill and H.J.C. Yeh, *J. Am. Chem. Soc.* **97**, 5602 (1975).
31. S.T. Hodgson, D.M. Hollinshead and S.V. Ley, *Tetrahedron* **41**, 5871 (1985).
32. G.D. Annis, E.M. Hebblethwaite, and S.V. Ley, *J. Chem. Soc. Chem. Commun.* 297 (1980).
33. A.M. Horton and S.V. Ley, *J. Organomet. Chem.* **285**, C17 (1985).
34. S.V. Ley, A. Armstrong, D. Diez-Martin, M.J. Ford, P. Grice, J.G. Knight, H.C. Kolb, A. Madin, C.A. Marby, S. Mukherjee, A.N. Shaw, A.M.Z. Slawin, S. Vile, A.D. White, D.J. Williams and M. Woods, *J. Chem. Soc. Perkin Trans.* **1**, 667 (1991).
35. J.I. Seeman and S.G. Davies, *J. Chem. Soc. Chem. Commun.* 1019 (1984).
36. S.G. Davies and J.I. Seeman, *Tetrahedron Lett.* **25**, 1845 (1984).
37. S.G. Davies, J.I. Seeman, and I.H. Williams, *Tetrahedron Lett.* **27**, 619 (1986).
38. S. L. Brown, S. G. Davies, D. F. Foster, J. I. Seeman and P. Warner, *Tetrahedron Lett.* **27**, 723 (1986).
39. B. K. Blackburn, S. G. Davies and M. Whittaker, *J. Chem. Soc. Chem. Commun.* 1344 (1987).
40. S. G. Davies, I. M. Dordor-Hedgecock, K. H. Sutton and M. Whittaker, *J. Am. Chem. Soc.* **109**, 5711 (1987).
41. L. S. Liebeskind, M. E. Welker and V. Goedken, *J. Am. Chem. Soc.* **106**, 441 (1984).
42. L. S. Liebeskind and M. E. Welker, *Organometallics* **2**, 194 (1983).
43. N. Aktogu, H. Felkin and S. G. Davies, *J. Chem. Soc. Chem. Commun.* 1303 (1982).
44. G. J. Baird, J. A. Bandy, S. G. Davies and K. Prout, *J. Chem. Soc. Chem. Commun.* 1202 (1983).
45. L. S. Liebeskind, R. W. Fengl and M. E. Welker, *Tetrahedron Lett.* **26**, 3075 (1985).
46. G. Bashiardes and S. G. Davies, *Tetrahedron Lett.* **28**, 5563 (1987).
47. S. G. Davies, I. M. Dordor and P. Walker, *J. Chem. Soc. Chem. Commun.* 956 (1984).
48. S. G. Davies, I. M. Dordor-Hedgecock, P. Warner, R. H. Jones and K. Prout, *J. Organomet. Chem.* **285**, 213 (1985).
49. S. G. Davies, I. M. Dordor-Hedgecock and P. Warner, *Tetrahedron Lett.* **26**, 2125 (1985).
50. S. G. Davies, G. Bashiardes, R. P. Beckett, S. J. Coote, I. M. Dordor-Hedgecock, C. L. Goodfellow, G. L. Gravatt, J. P. McNally and M. Whittaker, *Phil. Trans. R. Soc. Lond. A*, **326**, 619 (1988).
51. S. G. Davies and G. L. Gravatt, *J. Chem. Soc. Chem. Commun.* 870; 871 (1988).
52. K. Broadley and S. G. Davies, *Tetrahedron Lett.* **25**, 1743 (1984).
53. G. Bashiardes, S. P. Collingwood, S. G. Davies and S. C. Preston, *J. Chem. Soc. Perkin Trans.* **1**, 1162 (1989).
54. S. P. Collingwood, S. G. Davies and S. C. Preston, *Tetrahedron Lett.* **31**, 4067 (1990).
55. S. G. Davies and J. C. Walker, *J. Chem. Soc. Chem. Commun.* 209 (1985).
56. S. G. Davies, I. M. Dordor-Hedgecock, K. H. Sutton and J. C. Walker, *Tetrahedron* **42**, 5123 (1986).
57. P. W. Jolly and R. Pettit, *J. Am. Chem. Soc.* **88**, 5044 (1966).
58. M. Brookhart and G. O. Nelson, *J. Am. Chem. Soc.* **99**, 6099 (1977).
59. S. Brandt and P. Helquist, *J. Am. Chem. Soc.* **101**, 6473 (1979).
60. E. J. O'Connor, S. Brandt and P. Helquist, *J. Am. Chem. Soc.* **109**, 3739 (1987).
61. C. Knorrs, G-H. Kuo, J. W. Lauher, C. Eigenbrot and P. Helquist, *Organometallics* **6**, 988 (1987).
62. P. Seutet and P. Helquist, *Tetrahedron Lett.* **29**, 4921 (1988).
63. S-K. Zhao, C. Knorrs and P. Helquist, *J. Am. Chem. Soc.* **111**, 8527 (1989).
64. S-K. Zhao and P. Helquist, *J. Org. Chem.* **55**, 5820 (1990).
65. A. G. M. Barrett and M. A. Sturgess, *J. Org. Chem.* **52**, 3940 (1987).
66. A. G. M. Barrett, J. Mortier, M. Sabat and M. A. Sturgess, *Organometallics* **7**, 2553 (1988).
67. For a review of applications of organometallic complexes in β-lactam synthesis, see: A. G. M. Barrett and M. A. Sturgess, *Tetrahedron* **44**, 5615 (1988).
68. M.L.H. Green and P.L.I. Nagy, *J. Chem. Soc.* 189 (1963).
69. A. Cutler, D. Ehntholt, P. Lennon, K. Nicholas, D.F. Marten, M. Madhavarao, S. Raghu, A. Rosan and M. Rosenblum, *J. Am. Chem. Soc.* **97**, 3149 (1975).

70. G.D. Annis, S.V. Ley, C.R. Self and R. Sivaramakrishnan, *J. Chem. Soc. Perking Trans.* **1**, 270 (1981).
71. T.S. Piper and G. Wilkinson, *J. Inorg. Nucl. Chem.* **3**, 104 (1956).
72. J.B. Bibler and A. Wojcicki, *Inorg. Chem.* **5**, 889 (1966).
73. R.B. King and M.B. Bisnette, *Inorg. Chem.* **4**, 486 (1965).

–4–

Diene Complexes of Iron

A tricarbonyliron moiety ($Fe(CO)_3$) that is coordinated to a 1,3-diene has remarkable effects on the diene. It may act as a protecting group, preventing reactions normally associated with carbon–carbon double bonds (e.g., hydroboration, osmylation) and with 1,3-dienes (e.g., Diels–Alder reactions), and as an activating group, leading to capabilities for achieving transformations such as nucleophile addition that do not occur under normal circumstances. The reactivity of the diene towards electrophilic reagents is moderated, allowing greater selectivity during such reactions, and very reactive dienes are stabilized, so that the $Fe(CO)_3$ group can be used as a temporary stabilizing group for antiaromatic molecules such as cyclobutadiene and cyclopentadienone.

4.1 PREPARATION OF DIENE-$FE(CO)_3$ COMPLEXES

Butadieneiron tricarbonyl (**4.1**) was reported by Reihlen *et al.* in 1930[1]. It may be prepared by direct reaction of 1,3-butadiene with iron pentacarbonyl at elevated temperature (*ca* 130–140°C), but since butadiene is a gas at normal temperatures it is necessary to carry out this reaction in a closed system. Two carbonyl ligands are successively replaced by olefinic groups to give a complex in which all four diene carbons are attached to the metal. With less volatile dienes, the reaction may be conducted in an open system under nitrogen or argon bubbler and in high-boiling solvent, most commonly di-n-butyl ether. *(CAUTION: Iron pentacarbonyl is toxic and should be used only in an efficient fume hood; since carbon monoxide, also toxic, is evolved during this reaction it is essential to ensure that no outgassing occurs into the general laboratory area).* For the beginner in this area, a few useful hints follow. Iron pentacarbonyl tends to react slowly with oxygen that has entered the storage bottle during its use, and deposits of iron oxide are often found in the bottle. It is recommended that the iron pentacarbonyl be filtered through a plug of cotton wool directly into the reaction vessel prior to its use. Di-n-butyl ether forms peroxides on storage, and these may cause oxidation of the iron carbonyls during their preparation. Filtration of the solvent through basic alumina prior to use is normally sufficient to remove peroxides. The reactant solution should be purged with nitrogen or argon prior to heating, and the reaction should be conducted under inert atmosphere. During the course of the reaction, black insoluble material is formed. Do not be alarmed (yet), since this is metallic iron formed by thermal decomposition of the iron pentacarbonyl. *It is, however, finely divided and, therefore, pyrophoric*. At the end of the reaction, after cooling the mixture, the iron is removed by filtration through celite. Because of its pyrophoric nature, this must not be allowed to dry out on the filter pad. As long as it is kept wet with solvent during the filtration and washing procedure, this presents no problem and it can then be destroyed with dilute hydrochloric acid. Most of these preparations employ an excess of iron pentacarbonyl, and some will remain in the reaction mixture. Therefore, the subsequent manipulations during isolation of the complex must be carried out in the fume hood. Removal of solvent and excess iron pentacarbonyl can be effected using a rotary evaporator equipped with a solid CO_2/acetone condenser. The excess iron pentacarbonyl freezes onto the condenser, while the solvent is collected in the receiver. In this way, the residual iron carbonyl may be recovered (by allowing the condenser to warm and collecting the iron carbonyl in a suitable receiver) or destroyed (by allowing it to drip into bromine water).

$Fe(CO)_5$, Δ → $Fe(CO)_3$ **4.1** (4.1)

Alternative methods for the preparation of diene-$Fe(CO)_3$ complexes include treatment of the diene with nonacarbonyldiiron, [$Fe_2(CO)_9$], dodecacarbonyltriiron [$Fe_3(CO)_{12}$], and with benzylideneacetoneiron tricarbonyl [(bda)$Fe(CO)_3$]. The latter is useful as an $Fe(CO)_3$ transfer reagent, and modifications of the procedure have led to approaches for the preparation of optically active diene-$Fe(CO)_3$ complexes, as will be discussed in Chapter 5. It is also useful in cases where reactions of the diene with $Fe(CO)_5$ or $Fe_2(CO)_9$ fail to give satisfactory yields of complex. The use of nonacarbonyldiiron is limited to reactions with 1,3-dienes, but it can be used under milder conditions than $Fe(CO)_5$ (lower temperature) and is therefore useful for complexation of more sensitive dienes. It is more expensive than $Fe(CO)_5$ but is readily prepared by photolysis of the latter in acetic acid (see Chapter 1).

A large number of 1,4-cyclohexadienes are available from the Birch reduction of aromatic compounds[2,3]. Reaction of these dienes with $Fe(CO)_5$ at *ca* 140°C (refluxing in di-n-butyl ether) affords 1,3-cyclohexadiene-$Fe(CO)_3$ complexes, in which conjugation of the 1,4 diene has occurred. The main problem with this method is that mixtures of complexes are obtained from substituted cyclohexadienes. For example, dihydrotoluene (**4.2**) and dihydroanisole (**4.3**) give approximately equimolar mixtures of complexes as indicated in equation 4.2.[4] From a synthetic standpoint this is a considerable disadvantage, but in some cases a remedy lies in preconjugating the diene or subjecting the product mixture to conditions suitable for isomerization to the thermodynamically most stable isomer.

$$\text{4.2 R = Me; 4.3 R = OMe} \xrightarrow{Fe(CO)_5,\ Bu_2O,\ \Delta} \text{4.4 R = Me; 4.5 R = OMe} + \text{4.6 R = Me; 4.7 R = OMe} \tag{4.2}$$

Some examples are presented here, but the reader should bear in mind that extensive studies on the preparation of very highly substituted cyclohexadiene complexes have not been undertaken.

Dihydroanisoles are readily converted into thermodynamic mixtures of 1,4- and 1,3-dienes, favoring the latter, by treatment with a catalytic amount of *p*-toluenesulfonic acid at *ca* 80°C[5]. Since this conjugation proceeds via protonation of the vinyl ether double bond, it has obvious limitations, but nevertheless provides a useful entry into the preparation of isomerically pure diene-$Fe(CO)_3$ complexes in good yield. Treatment of the equilibrium mixture of dienes with iron pentacarbonyl under the usual conditions leads to a product which is predominantly the unrearranged 1,3-diene complex[6]. Purification by column chromatography on silica gel affords isomerically pure compounds, two examples being given in equation 4.3.

$$\text{4.3 R = H; 4.8 R = Me} \underset{}{\overset{p\text{-TsOH, 80 °C}}{\rightleftharpoons}} \text{4.9 R = H; 4.10 R = Me} \xrightarrow[\text{2) Chromatography}]{\text{1) }Fe(CO)_5,\ Bu_2O,\ \Delta} \text{4.7 R = H; 4.11 R = Me} \tag{4.3}$$

This method is amenable to scale-up. Indeed, **4.11** and related complexes have been used by the author's research group as intermediates in the synthesis of a number of natural products, and this is discussed in more detail in Chapter 5. Typically, the preparation of **4.11** on 200–300 g scale presents no problem in the research laboratory. Moreover, these complexes are relatively stable, both thermally and towards air, and they can be stored almost indefinitely in the freezer. Complexes that are oils tend to undergo slow aerobic oxidation and a precautionary measure of storing under nitrogen or argon is advisable.

Recent work has shown that stereoselectivity can be obtained during the complexation of cyclohexadiene diols. Discussion of this is deferred to Chapter 5.

The preceding discussion gives a general outline of methods for the preparation of diene-$Fe(CO)_3$ complexes. The following discussion will focus on reactions of a number of specific complexes in order to illustrate areas of utility. Where appropriate the method of preparation of the complex will be given.

4.2 USES OF TRICARBONYLIRON AS A DIENE PROTECTING GROUP

Lewis and coworkers[7] showed that the diene-Fe(CO)$_3$ group in tricarbonyl(myrcene)iron (**4.12**) and tricarbonyl(ergosteryl acetate)iron (**4.17**) is unreactive toward hydroboration, osmylation and hydration, whereas Barton *et al.*[8] also showed that hydrogenation of the 22,23-double bond of the ergosteryl benzoate complex (**4.18**) proceeded selectively without detriment to the diene-Fe(CO$_3$ moiety. These reactions are summarized in equations 4.4–4.9. Of course, the Fe(CO)$_3$ group is only useful as a diene protection if it can be introduced and removed in good yield and high selectivity. This has been accomplished in a large number of cases, one method of demetallation (ceric amonium nitrate) being shown in equations 4.4 and 4.5.

Fe(CO)$_3$ — AcOH, BF$_3$ → Fe(CO)$_3$ … OAc — Ce(IV) → OAc (4.4)

4.12 **4.13** **4.14**

4.12 — 1) BH$_3$ 2) MeOH → Fe(CO)$_3$ … B(OMe)$_2$ — Ce(IV) → B(OMe)$_2$ (4.5)

RO … H — (bda)Fe(CO)$_3$ or Fe$_2$(CO)$_9$ + MeOC$_6$H$_4$CH=CHCOCH$_3$ (>70%) → RO … H … Fe(CO)$_3$ (4.6)

4.17 R = Ac
4.18 R = PhCO

4.17 → (OsO_4, py; (85%)) 4.19 (OH, OH) (4.7)

4.17 → (1) BH_3.THF; 2) H_2O_2, NaOH) 4.20 (OH) + 4.21 (OH) (R = H) (4.8)

4.18 → (H_2, Pt, $Me_2PhCH_2SiH_2$; (94%)) 4.22 (4.9)

It should be noted that, owing to the environment of the 22,23 double bond in **4.17**, hydroboration and osmylation occur in a regio- and stereochemically random fashion. Greater control can be obtained during these reactions if a more appropriate substrate is chosen. For example, hydroboration and osmylation of η^4-cycloheptatriene-$Fe(CO)_2L$ (L = CO or $P(OPh)_3$) complexes proceed with complete regio- and stereoselectivity[9]. Perhaps the most interesting examples are those shown in equations 4.12 and 4.13 where the stereo- and regiodirecting effects of the $Fe(CO)_2P(OPh)_3$ group overcome steric hindrance from the methyl substituent. It is also noteworthy that *catalytic* osmylation of **4.24** can be accomplished, since one may not have expected the diene-$Fe(CO)_3$ to withstand treatment with the co-oxidant. This has great practical advantage over the stoichiometric method, owing to the high cost of osmium tetraoxide. The regiochemical outcome during hydroboration of **4.24** and **4.27** is explained by the pronounced ability of the diene-$Fe(CO)_2L$ to stabilize positive charge on a proximal carbon atom, as would be encountered in the polar transition state for addition of -B-H to the uncomplexed olefinic group. This effect is related to the facility with which diene- and triene-$Fe(CO)_2L$ complexes can be converted into cationic dienyl-$Fe(CO)_2L$ systems, and is discussed in more detail in Chapter 5.

$Fe(CO)_3$ 1) $NaBH_4$, $CeCl_3$ 2) TBSOTf, py, CH_2Cl_2 $Fe(CO)_3$ OTBS OsO_4; $NaHSO_3$ (98%) or Bu^tOOH, cat. OsO_4 Et_4NOAc (70%) $Fe(CO)_3$ OTBS HO OH (4.10)

4.23 **4.24** (98%) **4.25**

4.24 BH_3.THF; H_2O_2, NaOH $Fe(CO)_3$ OTBS OH (4.11)

4.26 (90%)

$Fe(CO)_2P(OPh)_3$ Me OsO_4; $NaHSO_3$ $Fe(CO)_2P(OPh)_3$ Me HO OH (4.12)

4.27 **4.28** (70 – 80%)

4.27 BH_3.THF; H_2O_2, NaOH $Fe(CO)_2P(OPh)_3$ Me OH (4.13)

4.29 (96%)

Considerable advantage has been taken of the ability of the $Fe(CO)_3$ to protect dienes and latent enone functionality during total syntheses of some natural products. Details of these syntheses will be given in Chapter 5, since they illustrate the tactics involved in employing reactive cyclohexadienyliron complexes in the key carbon–carbon bond forming reactions. Some examples of osmylation[6,10] and even Sharpless epoxidation[11,12] are shown in equations 4.14 and 4.15. It is unlikely that these reactions could have been carried out as efficiently subsequent to decomplexation and unmasking of the cyclohexenone functionality (from acidic hydrolysis of the dienol ether).

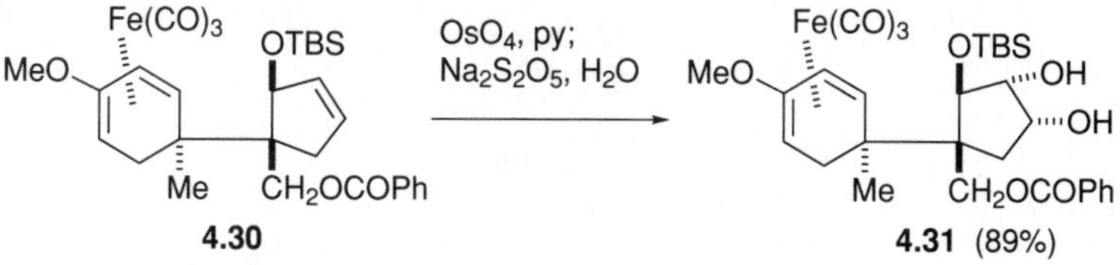

(4.14)

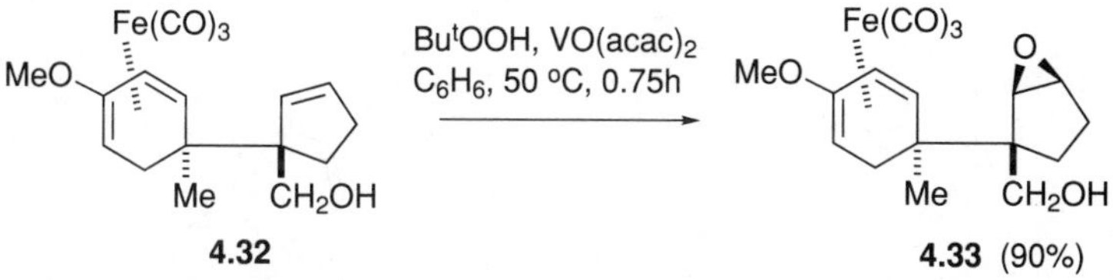

(4.15)

Carbon–carbon bond forming reactions on the uncomplexed double bond of cycloheptatriene-$Fe(CO)_3$ complexes have also been described. Cyclopropanation of **4.23**, using diazoethane[13], affords the complex **4.34**, which is readily demetallated using trimethylamine-*N*-oxide[14]. Reaction of complex **4.36** with electrophilic acylating agents occurs with complete regioselectivity to give **4.37** and **4.38**, respectively[15]. Diels–Alder reaction of the uncomplexed double bond has also been achieved under high pressure conditions[16]. This behavior has been extended to the manipulation of tropone-$Fe(CO)_3$ (**4.23**) to give the natural products β-thujaplicin (**4.41**) and β-dolabrin (**4.42**)[17,18].

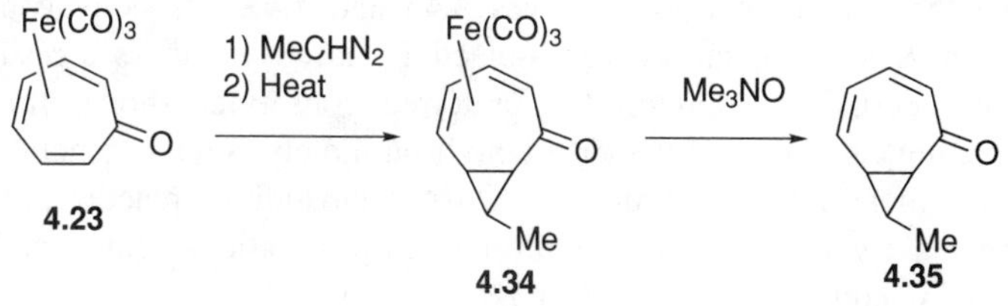

(4.16)

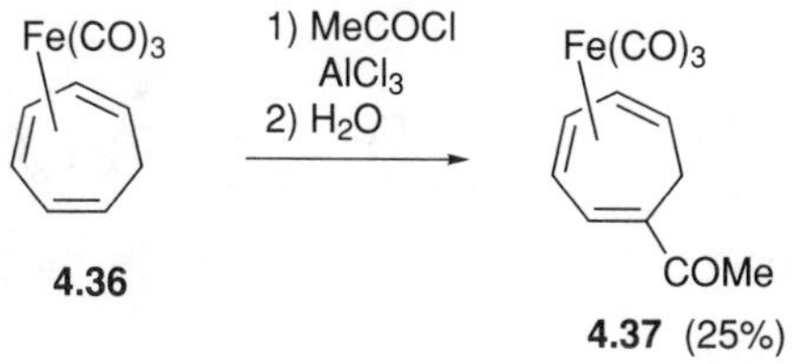

(4.17)

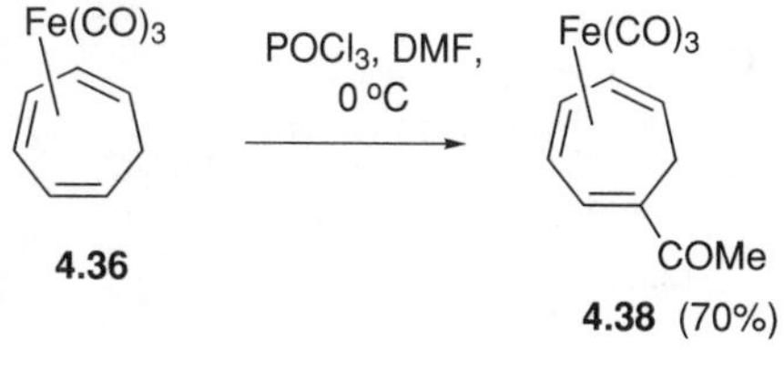

(4.18)

1) MeCOCl AlCl$_3$ 2) H$^+$; 1) Me$_2$CHN$_2$ 2) CH$_2$Cl$_2$, Δ

4.23 → **4.39** (80%) → **4.40** (83%) → **4.41**, **4.42**

Fe(CO)$_3$ Me Me$_2$CH X R

4.41 R = CHMe$_2$, X = OH
4.42 R = MeC=CH$_2$, X = H

(4.19)

The above reaction types have been explored in reactions with butadiene-Fe(CO)$_3$ complexes having pendant olefinic groups, and this chemistry has been reviewed[19]. A number of cycloaddition reactions on the uncomplexed double bond proceed cleanly to give cyclopropanes, cyclobutanes and heterocycles of various types. Dichlorocarbene addition to myrcene-Fe(CO)$_3$ (**4.12**) and complex **4.44** give the cyclopropane derivatives **4.43** and **4.45**, respectively. During the reactions of **4.44**, complete diastereoselectivity is observed, as a result of attack *anti* to the Fe(CO)$_3$ group on the preferred conformer shown for **4.44**. The stereodirecting capability of the tricarbonyliron moiety is quite general and will be discussed in more detail in Section 4.3. Other cycloaddition reactions of the acyclic complexes, as well as Michael additions, hydroborations and osmylations, are summarized in equations 4.22–4.26.

:CCl$_2$

4.12 Fe(CO)$_3$ → **4.43** Fe(CO)$_3$ Cl Cl

(4.20)

:CCl$_2$, phase transfer conditions

4.44 Fe(CO)$_3$ MeO$_2$C H Me Me → **4.45** Fe(CO)$_3$ MeO$_2$C H Cl Cl Me Me

(4.21)

$Fe(CO)_3$ RCO 4.46 Me $\xrightarrow{(NC)_2C{=}C(CN)_2}$ $Fe(CO)_3$ RCO 4.47 CN CN CN CN Me (4.22)

$Fe(CO)_3$ MeO_2C H CO_2Me CO_2Me 4.48 $\xrightarrow{CH_2N_2}$ $Fe(CO)_3$ MeO_2C H N N CO_2Me CO_2Me 4.49

CAN

MeO_2C H CO_2Me CO_2Me 4.50 (4.23)

$Fe(CO)_3$ MeO_2C H CO_2Me CO_2Me 4.48

$Me_2S{=}CHCO_2Me$

$Fe(CO)_3$ MeO_2C H CO_2Me CO_2Me CO_2Me 4.51 (4.24)

$Fe(CO)_3$ 4.52 $\xrightarrow{\text{1) } BH_3 \text{ 2) } H_2O_2\text{, NaOH}}$ $Fe(CO)_3$ OH 4.53 (4.25)

$Fe(CO)_3$ 4.52 $\xrightarrow{OsO_4\text{, etc.}}$ $Fe(CO)_3$ OH OH 4.54 (4.26)

The preceding discussion suggests that diene-$Fe(CO)_3$ complexes are completely unreactive to reagents that normally attack olefin systems. This is not true for all such reagents. For example, Friedel–Crafts acylation[20] of complex **4.12** does indeed proceed selectively on the uncomplexed double bond to give complex **4.55** and this appears to suggest that no acylation occurs with diene-$Fe(CO)_3$ moieties. However, butadiene-$Fe(CO)_3$ (**4.1**) does react with $CH_3COCl/AlCl_3$ to give acyl-substituted complexes **4.56**[21,22].

$Fe(CO)_3$ **4.12** — 1) $MeCOCl/AlCl_3$, 2) H_2O → $Fe(CO)_3$, COMe **4.55** (4.27)

$Fe(CO)_3$ **4.1** — 1) $MeCOCl/AlCl_3$, 2) H_2O → $Fe(CO)_3$, COMe **4.56** (4.28)

It should be noted that the $Fe(CO)_3$ group does moderate the reactivity of the diene, and in fact leads to considerable control, since under Friedel–Crafts conditions 1,3-dienes normally undergo facile polymerization reactions[23], which are prevented by the attachment of the iron moiety. The acylation of **4.1** has been shown to proceed *via* the intermediacy of a π-allyl-$Fe(CO)_3$ complex such as **4.57,** in which the vacant coordination site on the metal is filled by donation from the acetyl oxygen lone pair[21,22]. Kinetic deprotonation (using cold aqueous ammonia) leads to the complex **4.58**, whereas work-up under equilibrating conditions gives exclusively **4.59**, thus providing some measure of stereocontrol in these reactions. Interestingly, the 2-methoxybutadiene complex **4.60** gives **4.61** as a single regioisomer[24], attributable to the effect of back-donation from iron into the diene LUMO[25]. It should be noted that the free diene is expected to react with electrophiles at C-1 and that this reversal of selectivity compared with uncomplexed ligands is quite common in organometallic chemistry.

$[(CO)_3Fe \leftarrow O{=}C(Me)]^+$ **4.57**

$Fe(CO)_3$, H, COMe **4.58**

$Fe(CO)_3$, COMe, H **4.59**

MeO, $Fe(CO)_3$ **4.60** — 1) $MeCOCl/AlCl_3$, 2) H_2O → MeO, $Fe(CO)_3$, H, COMe **4.61** (4.29)

Cyclohexadiene-Fe(CO)$_3$ (**4.62**) does undergo Friedel–Crafts acylation[26,27] but the yields are low. This may be remedied by replacing one CO ligand by PPh$_3$, leading to a more electron-rich system[28]. Reaction of the derived complex **4.63** with CH$_3$COCl/AlCl$_3$ proceeds cleanly in methylene chloride at −78°C to give the *endo* acetyl complex **4.64**. The stereoselectivity observed for this reaction suggests that acylation occurs first on the metal and the acetyl group is transferred to the diene in a subsequent step.

Fe(CO)$_2$L 1) MeCOCl/AlCl$_3$ CH$_2$Cl$_2$, −78°C 2) H$_2$O Fe(CO)$_2$L COMe H (4.30)

4.62 L = CO PPh$_3$, Δ
4.63 L = PPh$_3$
4.64 L = PPh$_3$ (96%)

Under appropriate conditions diene-Fe(CO)$_3$ complexes can be made to react with olefins. With the earliest examples[29–34] it was found that under photochemical conditions complexes such as **4.1** and **4.62** react with electron-deficient alkenes to give π-allyl complexes such as **4.65** (equation 4.31). The photochemical conditions are required for ejection of CO from the complex to create a vacant coordination site which is then occupied by the attacking alkene. Subsequent coupling, followed by reattachment of CO gives the product complex. As it stands this reaction is somewhat limited, because of the requirement for highly electron-deficient alkenes and because the product π-allyl complex has not been converted into useful organic molecules. A modification of this reaction, which involves thermal intramolecular coupling of alkene and diene followed by hydrogen migration and reductive elimination, has been described[35]. The probable mechanism of this reaction is shown in Scheme 4.1 and several examples are given in equations 4.32–4.35. The yields are all high, provided the reaction is carried out under carbon monoxide atmosphere (required for the conversion of electron-deficient diene-Fe(CO)$_2$ complex **4.71** to the product **4.72**). These reactions are quite unique in their ability to couple highly substituted alkenes giving product molecules having contiguous quaternary centers though they have not yet been applied in, e.g., natural products synthesis.

Fe(CO)$_3$ CF$_2$=CF$_2$ *hν* (CO)$_3$Fe F F F F (4.31)

4.62 **4.65**

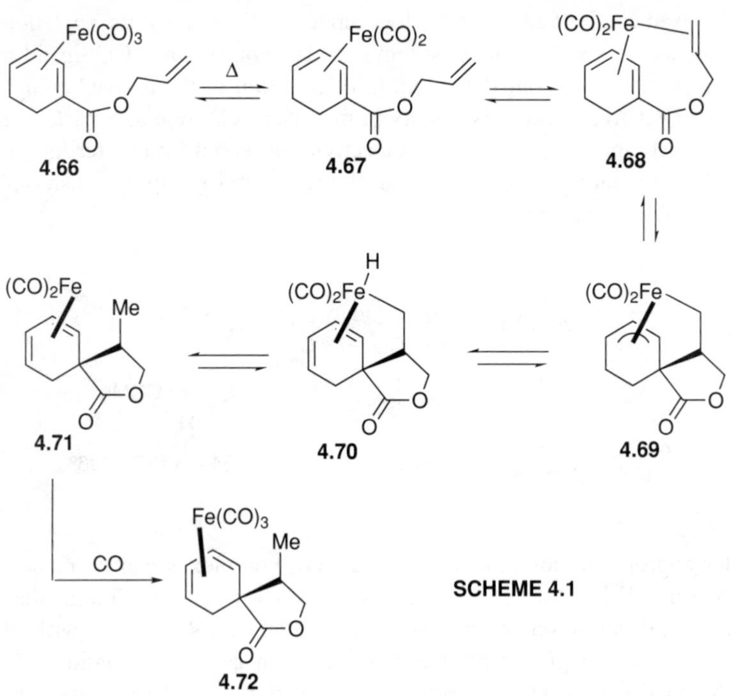

Fe(CO)3 Δ Fe(CO)2 (CO)2Fe

4.66 4.67 4.68

(CO)2Fe Me (CO)2Fe H (CO)2Fe

4.71 4.70 4.69

CO Fe(CO)3 Me

4.72

SCHEME 4.1

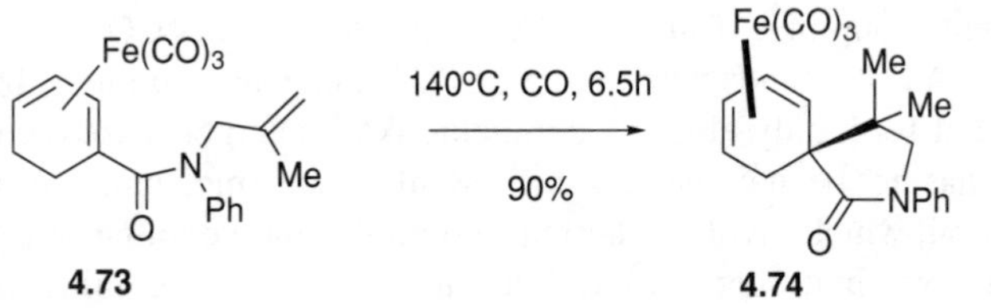

Fe(CO)3 140ºC, CO, 6.5h 90% Fe(CO)3 Me Me N Ph NPh (4.32)

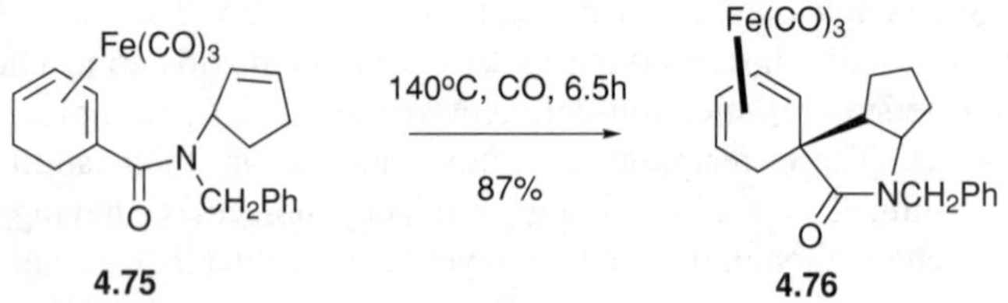

4.73 4.74

Fe(CO)3 140ºC, CO, 6.5h 87% Fe(CO)3 N CH2Ph NCH2Ph (4.33)

4.75 4.76

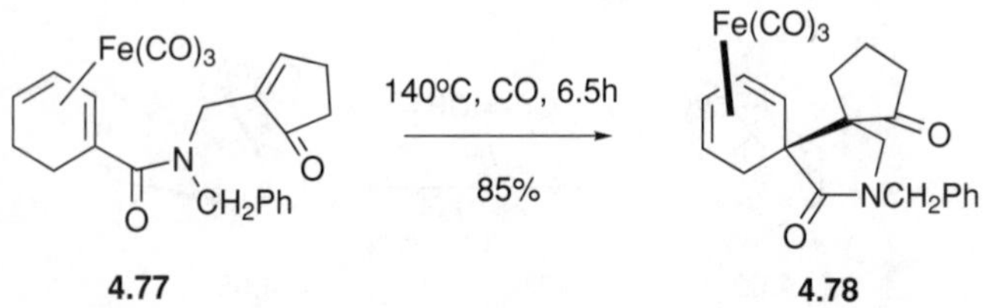

Fe(CO)3 140ºC, CO, 6.5h 85% Fe(CO)3 N CH2Ph NCH2Ph (4.34)

4.77 4.78

$Fe(CO)_3$ N CH_2Ph O **4.79** — 140°C, CO, 6.5h / 58% → $Fe(CO)_3$ NCH_2Ph O **4.80** (4.35)

Takacs and Anderson[36] have developed some related coupling reactions of dienes and olefins that are promoted by iron(0), though the outcome is somewhat different (equation 4.36).

OCH_2Ph **4.81** — Fe(0), bipy / 25°C, PhH, 3h → OCH_2Ph **4.82** (mixture of double bond isomers) (4.36)

4.3 USES OF TRICARBONYLIRON AS A STEREOCHEMICAL CONTROLLER

Some examples of this property of the organoiron moiety have already been encountered (see equations 4.10–4.13 and 4.21). Stereocontrol during reactions of functionality proximal to the diene-$Fe(CO)_3$ system depends on the steric bulk of the $Fe(CO)_3$ group. An interesting application of this concept is given in the work of Barton *et al.*[8] who converted the ergosterol complex **4.83** to *epi*-ergosterol. Oxidation of **4.83** using the Corey procedure gave the ketone **4.84**, in which the olefinic bonds were prevented from becoming conjugated with the ketone carbonyl by their complexation to iron. Reduction of **4.84** with the bulky reducing agent $LiAl(OBu^t)_3H$ occurred from the normally hindered β-face of the steroid, as a result of the $Fe(CO)_3$ group blocking the α-face, and demetallation of the resulting complex gave the previously unknown *epi*-ergosterol (**4.85**).

The directing ability of the $Fe(CO)_3$ group, coupled with conformational preferences of neighboring functionality, has been utilized[19] to effect stereocontrol in acyclic systems. Coupled with resolution of the chiral starting complexes, this methodology provides access to optically pure organic compounds. For example, resolution of the aldehyde **4.86** can be accomplished via formation of the oxazolidines **4.87**, followed by chromatographic separation and conversion into the individual enantiomerically pure aldehydes. Reaction of the aldehyde with nucleophiles proceeds with moderate to very good diastereoselectivity, depending

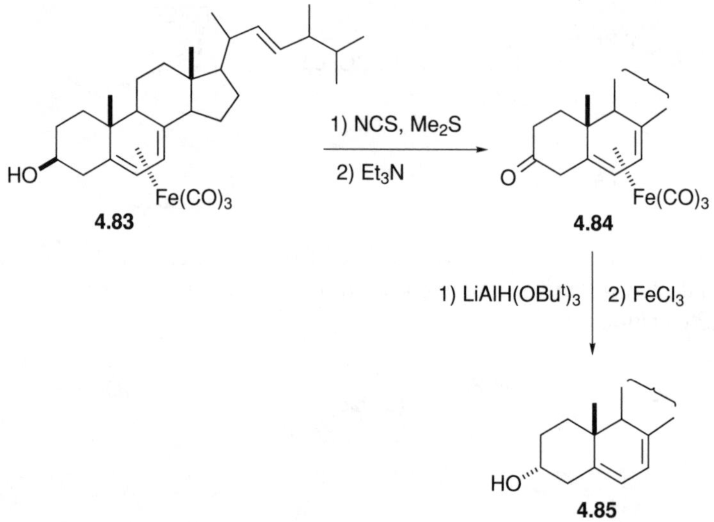

HO
Fe(CO)3
4.83
1) NCS, Me2S
2) Et3N
O
Fe(CO)3
4.84
1) LiAlH(OBut)3
2) FeCl3
HO
4.85

(4.37)

on the nucleophile. These reactions appear to occur by nucleophile attack *anti* to the metal with the organic ligand in s-*cis* or s-*trans* conformation, the relative population of which depends on the nature of the nucleophile and the reaction conditions. Some examples of reactions of **4.86** are shown in equations 4.38–4.40.

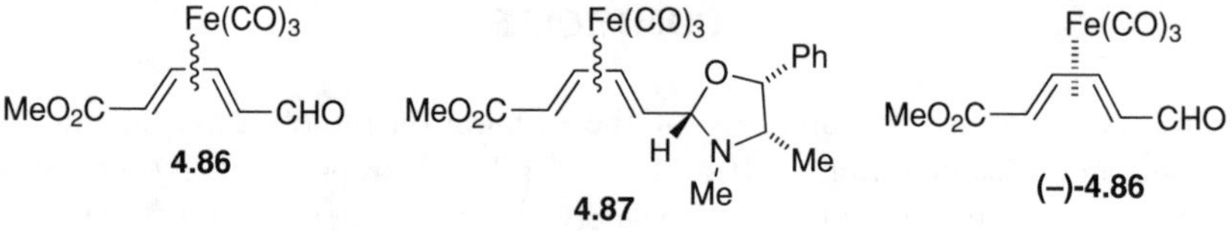

Fe(CO)3
MeO2C
CHO
4.86
Fe(CO)3
MeO2C
O
Ph
H
N
Me
Me
4.87
Fe(CO)3
MeO2C
CHO
(–)-4.86

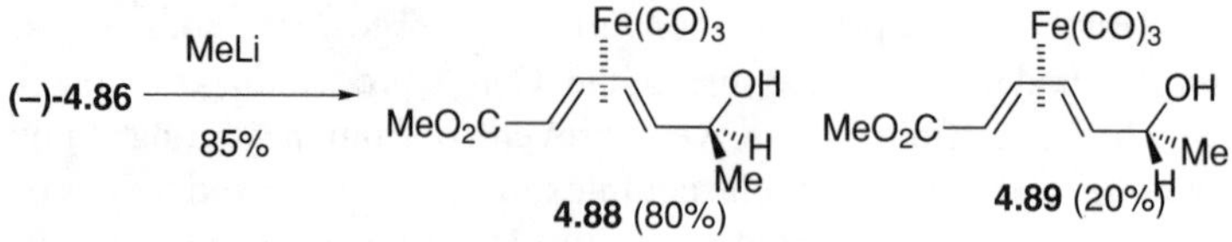

(–)-4.86
MeLi
85%
Fe(CO)3
MeO2C
OH
H
Me
4.88 (80%)
Fe(CO)3
MeO2C
OH
Me
H
4.89 (20%)

(4.38)

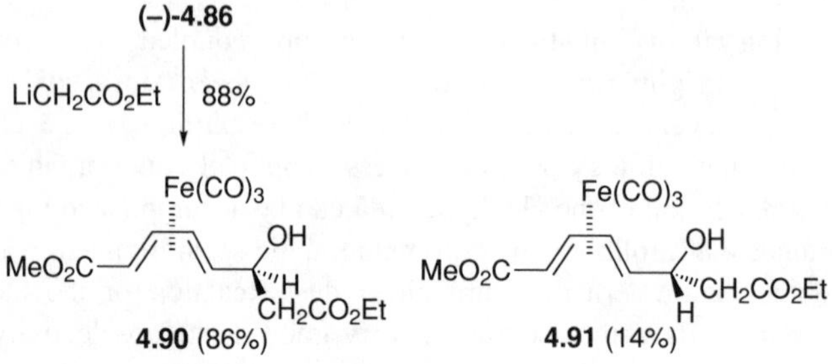

(–)-4.86
LiCH2CO2Et
88%
Fe(CO)3
MeO2C
OH
H
CH2CO2Et
4.90 (86%)
Fe(CO)3
MeO2C
OH
CH2CO2Et
H
4.91 (14%)

(4.39)

(–)-**4.86** —[$MeTi(OPr^i)_3$; 94%]→ $Fe(CO)_3$ MeO_2C OH H Me **4.88** (25%) + $Fe(CO)_3$ MeO_2C OH Me H **4.89** (75%) (4.40)

A very interesting example of kinetic resolution during the reaction of racemic **4.86** with optically pure chiral allylboronate **4.92** has been reported[37]. During this reaction it was found that the (2*R*) isomer reacts much faster with **4.92**, giving 92.5:7.5 selectivity in favor of the formation of diastereomer **4.93**, which was obtained in 96% enantiomeric excess. This excellent selectivity was extended to the reaction of (*R*,*R*)-**4.92** with the meso complex **4.94** to give a 45:1 mixture of complexes **4.95** and **4.96**, from which the major product was isolated in 82% yield and >98% enantiomeric excess. These results compare very favorably with the results of allylboronate additions to uncomplexed dienals which give much poorer stereoselectivity.

4.86 + (*S*,*S*)-**4.92** (O, B, O, CO_2Pr^i, CO_2Pr^i) —[toluene; –78°C, mol sieves]→ $Fe(CO)_3$ MeO_2C OH **4.93** (4.41)

$Fe(CO)_3$ OHC CHO **4.94** —[(*R*,*R*)-**4.92**]→ $Fe(CO)_3$ OHC OH **4.95** + $Fe(CO)_3$ OHC OH **4.96** (4.42)

(–)-**4.86** —[R-substituted allene]→ $Fe(CO)_3$ MeO_2C OH R → MeO_2C $OSiPh_2Bu^t$ R (4.43)

Combination of the above type of stereoselectivity and diene protection has led to the preparation of a key intermediate for the synthesis of leukotriene B (equation 4.43)[19].

4.4 ACTIVATION OF DIENES TOWARD NUCLEOPHILE ADDITION

Zerovalent transition metals that are coordinated to olefinic ligands have a tendency to act as electron acceptors and this leads to varying degrees of activation toward nucleophile addition to the alkene. Provided that the resulting complex can be converted into organic products, this provides a useful method for carbon–carbon bond formation. Of course, metals in zero oxidation state are less effective activators than higher valent species, which usually confer a positive charge on the resultant complex, but different regiochemical effects are often seen according to whether the attached metal is electron-rich or -poor[38]. This is certainly the case with diene-$Fe(CO)_3$ complexes, which behave very differently from the analogous positively charged diene-$Mo(CO)_2Cp$ systems[39], although they have not been extensively applied in synthesis. The results described here are taken from refs. [40] and [41].

Reaction of butadiene-$Fe(CO)_3$ (**4.1**) with 2-lithio-isobutyronitrile, followed by acid quench, gives the four olefinic products shown in equation 4.44. The major product appears to result from addition of the nucleophile at C-2, followed by acid-promoted demetallation.

$Fe(CO)_3$

4.1

1) $LiCMe_2CN$ 2) H^+

CMe_2CN CMe_2CN CMe_2CN CMe_2CN (4.44)

4.97 (89%) **4.98** (4%) **4.99** (6%) **4.100** (1%)

A very similar state of affairs is observed in the reactions of cyclohexadiene-$Fe(CO)_3$ (**4.62**). Mere inspection of the product structures from nucleophile addition followed by acid treatment (equation 4.45) does not allow us to determine whether attack occurs at C-2 or C-1 of the diene. However, if the reaction mixture is allowed to warm to room temperature before acid treatment an additional product (**4.104**, equation 4.46) is observed, which appears to result from CO insertion on a C-2 nucleophile addition product. Thus, the norm in most of these reactions seems to be nucleophilic attack at C-2 of the diene.

$$\textbf{4.62}\ [\text{Fe(CO)}_3] \xrightarrow[\text{2) H}^+,\ 98\%\ \text{yield}]{\text{1) LiCMe}_2\text{CN},\ -78^\circ\text{C}} \textbf{4.101}\ (87\%) + \textbf{4.102}\ (9\%) + \textbf{4.103}\ (4\%) \quad (4.45)$$

(each product bearing CMe_2CN)

$$\textbf{4.62} \xrightarrow[\text{2) H}^+]{\text{1) LiCMe}_2\text{CN},\ -78^\circ\text{C to }25^\circ\text{C}} \textbf{4.101}, \textbf{4.102}, \textbf{4.103} + \textbf{4.104} \quad (4.46)$$

(4.104 bearing CMe_2CN and CHO)

Carbonyl insertion can also be accomplished during the reactions of acyclic diene complexes, but here the reaction may follow a slightly different course depending on the diene, giving cyclopentanones via CO insertion coupled with a reductive elimination (equation 4.47 and 4.48)[42].

$$\textbf{4.1} \xrightarrow[\text{2) CF}_3\text{CO}_2\text{H},\ -78^\circ\text{C};\ 85\%]{\text{1) LiCMe}_2\text{CN, CO},\ -78^\circ\text{C to }25^\circ\text{C}} \textbf{4.105} \quad (4.47)$$

(4.105 bearing CMe_2CN and O)

$$\text{Me, Me diene-Fe(CO)}_3 \xrightarrow[\text{2) CF}_3\text{CO}_2\text{H},\ -78^\circ\text{C}]{\text{1) LiCMe}_2\text{CN, CO},\ -78^\circ\text{C to }25^\circ\text{C}} \textbf{4.107} \quad (4.48)$$

(4.107 bearing Me, Me, CMe_2CN and CHO)

4.5 STABILIZATION OF REACTIVE DIENES

The classic example of a reactive diene that can be stabilized by coordination to a metal is cyclobutadiene. Longuet-Higgins and Orgel[43] predicted that this antiaromatic compound could be stabilized as an organometallic complex, and there was subsequently much interest in the area. The first synthesis of tricarbonylcyclobutadieneiron (**4.108**) was reported in 1965[44] (equation 4.49). The complex acts as a convenient source of cyclobutadiene, which can be liberated and trapped as its cycloaddition products (equation 4.50). One application of this methodology is in an early synthesis of cubane derivatives (equation 4.51)[45–47].

(4.49)

4.108

(4.50)

4.109

(4.51)

4.110 **4.111**

Another antiaromatic molecule that can be stabilized by metal coordination is cyclopentadienone. A number of substituted cyclopentadienone-$Fe(CO)_3$complexes have been prepared by reaction of acetylenes with iron pentacarbonyl, although the yields are often low (equation 4.52)[48,49]. Some improvement was noted when the coupling reaction was carried out intramolecularly (equation 4.53)[50,51], but only recently has it been noted that moderate CO pressure during the intramolecular reaction gives excellent yields of the cyclopentadienone complexes (equation 4.54)[52,53]. With this modification we can anticipate the development of efficient methodology for the construction of di- and triquinanes, which are important components of a fairly large number of important natural product molecules[54–59].

(4.52)

4.112

(4.53)

4.113

X R Fe(CO)$_5$, CO 100 psi, toluene, 125–130°C X R =O Fe(CO)$_3$ R R (4.54)

4.114 X = H, R = Ph (81%)
4.115 X = H, R = $SiMe_3$ (85%)
4.116 X = OH, R = Ph (95%)

Another interesting and potentially useful reactive diene, which is stabilized by coordination to $Fe(CO)_3$ is cyclohexadienone, a tautomer of phenol. The complex (**4.118**) is prepared by hydrolysis of the methoxy-substituted cyclohexadienyl-$Fe(CO)_3$ complex **4.117** (see Chapter 5)[60]. The ketone carbonyl of **4.118** undergoes reaction with a limited range of nucleophiles and the products can be converted into substituted cyclohexadienyl complexes by treatment with acid. A modification of this method has recently been used for the preparation of aryl-substituted dienyl complexes that are useful precursors for synthesis of Amaryllidaceae alkaloids[61], discussed more fully in Chapter 5 (see equation 4.59 below).

OMe + Fe(CO)$_3$ BF_4^- **4.117** $\xrightarrow{H^+, H_2O, \Delta}$ O Fe(CO)$_3$ **4.118** (4.55)

4.118 $\xrightarrow{NABH_4}$ OH Fe(CO)$_3$ **4.119** (4.56)

4.118 $\xrightarrow{BrZnCH_2CO_2Et}$ HO CH_2CO_2Et Fe(CO)$_3$ **4.120** $\longrightarrow$ $CHCO_2Et$ Fe(CO)$_3$ **4.121** (4.57)

4.121 $\xrightarrow{HBF_4}$ CH_2CO_2Et + Fe(CO)$_3$ BF_4^- **4.122** (4.58)

OMe PF$_6^-$ Fe(CO)$_3$ OMe **4.123** ArLi MeO Ar Fe(CO)$_3$ MeO **4.124** (Ar = 4-MeOC$_6$H$_4$) 1) H$^+$ 2) NH$_4$PF$_6$ Ar Fe(CO)$_3$ PF$_6^-$ OMe **4.125** (4.59)

4.6 HETERODIENE COMPLEXES

Although heterodiene-Fe(CO)$_3$ complexes have been known for some time[62], they have only recently been afforded some attention with regard to synthetic application. In addition to the usual aza- and oxadiene complexes, the more recently developed vinyl ketene derivatives which are derived from enone-Fe(CO)$_3$ complexes will be included in this section. This discussion will be rather brief owing to the, as yet limited, number of real synthetic applications of these complexes.

The preparation of benzylideneacetone-Fe(CO)$_3$ (**4.126**) was reported in 1972[62], and has been used as an Fe(CO)$_3$ transfer reagent for the preparation of other diene-Fe(CO)$_3$ complexes (see p. 68). The reactions of heterodiene complexes with electrophiles, e.g., acylating agents, proceed as expected (equation 4.60). Only more recently have the reactions of heterodiene complexes with nucleophiles been studied[63] leading to interesting and potentially useful transformations.

Ph R O Fe$_2$(CO)$_9$ Fe(CO)$_3$ Ph R O **4.126** R = Me **4.127** R = H MeCOBF$_4$ Ph R O (CO)$_3$Fe O Me BF$_4^-$ **4.128** (4.60)

Thomas found that better yields (*ca* 80%) of the heterodiene complexes are obtained when an excess (2 equivalents) of Fe$_2$(CO)$_9$ is used. Reaction of enone-Fe(CO)$_3$ complexes with alkyllithium or Grignard reagents *under nitrogen atmosphere*, followed by protonation using t-butyl bromide, leads to the formation of 1,4-diketones, via addition of the nucleophiles to a CO ligand (equation 4.61)[64]. In contrast, when this reaction is carried out under *carbon monoxide atmosphere* vinylketene complexes are formed (equation 4.62)[65]. The corresponding azadiene-$Fe(CO)_3$ complex (**4.134**) undergoes reaction with alkyllithium reagents under N_2 atmosphere to give substituted pyrroles (equation 4.63), presumably via the keto imine corresponding to **4.132**, except in the case of complexes lacking a 2-alkyl substituent (equation 4.64)[66].

The vinylketene complexes **4.133** also undergo some interesting transformations. They are converted into vinylketenimine complexes upon reaction with isonitriles,

R^2Li, Et_2O, –78°C; **4.129** → [**4.130**] → [**4.131**] → Bu^tBr → **4.132** (4.61)

$R^1 = R^2 = Me$ (70–79%)
$R^1 = Me$, $R^2 = Bu^n$ (53%)
$R^1 = Bu^n$, $R^2 = Me$ (73%)

4.129 → 1) MeLi, THF, CO atmos. –78°C to rt; 2) Bu^tBr → **4.133** (4.62)

4.133 $R^1 = Me$ (35%)
$R^1 = Bu^t$ (83%)

4.134 → 1) MeLi, THF, CO atmos. –78°C, 6.5h; 2) Bu^tBr; 70% → **4.135** (4.63)

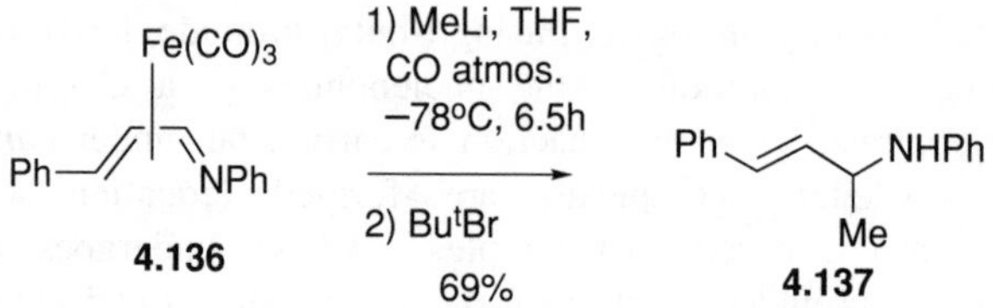

(4.64)

a transformation which has been shown to proceed via initial ligand exchange (equation 4.65)[67]. The derived vinylketenimine complexes behave differently from the parent vinylketene complexes on reaction with nucleophiles[68], as illustrated in equations 4.66 and 4.67.

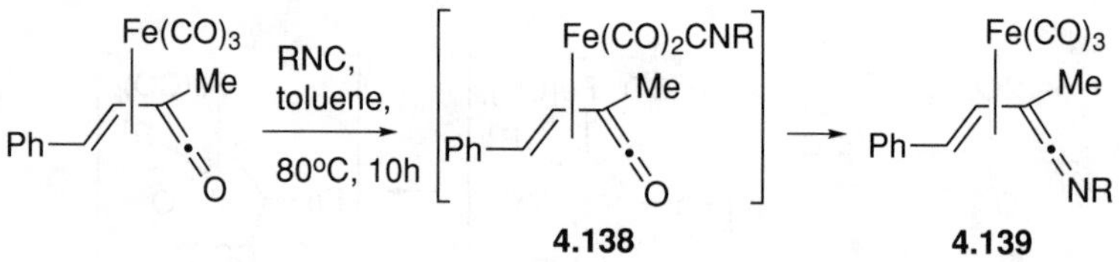

(4.65)

R = Bu^t (64%)
R = C_6H_{11} (59%)

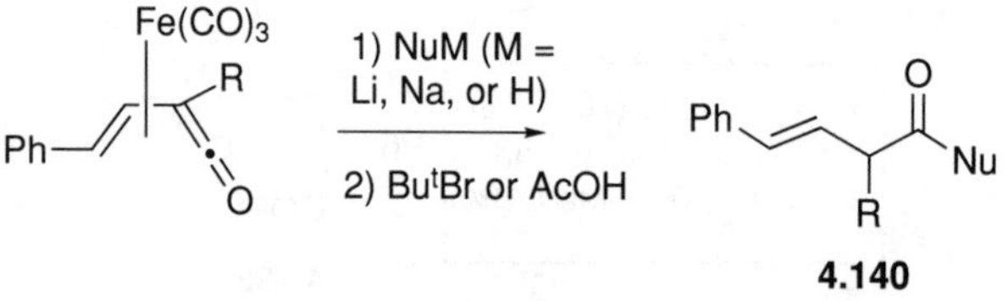

(4.66)

R = Me, Nu = Me (71%)
R = Me, Nu = CMe_2CO_2Et (67%)
R = Bu^t, Nu = $NHCH_2Ph$ (85%)
R = Bu^t, Nu = OMe (89%)

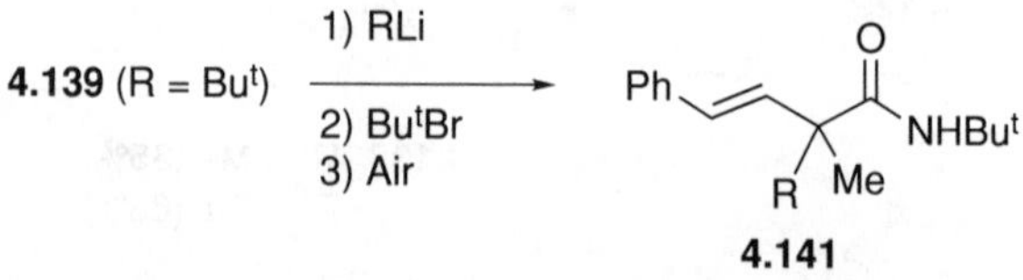

(4.67)

R = Me (53%)
R = Et (54%)
R = Bu^n (57%)
R = Bu^t (63%)

During some of these reactions (equations 4.64, 4.66 and 4.67) a chiral center is produced in the organic product. Since the complexes are themselves chiral there is a possibility that optically pure complexes would lead to optically pure organic products. With this in mind, the diastereoselectivity has been studied during complexation of azadienes and oxadienes bearing chiral auxiliaries. The most promising results are shown in equations 4.68[69] and 4.69[70].

$Fe_2(CO)_9$, 28% yield, 94:6 diastereoselectivity → **4.142** (4.68)

$(bda)Fe(CO)_3$ 1.4 equiv. → **4.143** (4.69)

R = Bu^t (67%)
R = Ph (64%) } single diastereomer

4.7 EXPERIMENTAL PROCEDURES

Preparation of tricarbonyl(1-methoxy-4-methylcyclohexa-1,3-diene)iron **(4.11)**

This procedure[6] represents a fairly typical method for the conversion of dienes into their $Fe(CO)_3$ complexes by reaction with pentacarbonyliron. All operations described for the preparation should be carried out in a fume hood. Rubber gloves should be worn when measuring and transferring pentacarbonyliron which is toxic and is absorbed by the skin. Birch reduction of *p*-methylanisole gives 1-methoxy-4-methylcyclohexa-1,4-diene which is partially conjugated by stirring with a catalytic amount of *p*-toluenesulfonic acid (100 mg 100 g^{-1} of diene) at 80°C under nitrogen for 2 h, followed by distillation under reduced pressure. A mixture of 1,4- and 1,3-dienes (1:3, 93% recovery) is obtained. This mixture (93.0 g) is dissolved in di-n-butyl ether (700 ml), purged with nitrogen and placed in a 2 liter three-neck flask fitted with a reflux condenser and overhead stirrer and placed in an electric heating mantle or an oil bath. The reflux condenser is equipped with a nitrogen bubbler (across the top; do not blow nitrogen through the flask since this will result in loss of pentacarbonyliron to the atmosphere). The apparatus is assembled in the fume hood (carbon monoxide is evolved). Pentacarbonyliron (250 ml) is added via the open third neck of the flask, with back flushing of nitrogen. The flask is then stoppered and the contents are stirred briskly while they are heated to gentle reflux. If an oil bath is used as the heat source a temperature of *ca* 145–150°C should be maintained; this procedure is more useful for small-scale preparations (up to 1 liter flask size). Heating is continued for 48 h, usually giving an ugly black mixture in the reaction vessel. The mixture is cooled to room temperature, allowing the black precipitate of (pyrophoric) iron to settle, and is vacuum filtered through a pad of Celite on a glass frit. It is best to decant the supernatant onto the Celite. Do not allow the filter pad to run dry, otherwise the pyrophoric iron can ignite; this can be easily controlled by manipulating the aspirator pressure in the usual way. The residue in the flask is washed by decantation several times with di-n-butyl ether and the washings are filtered through the Celite pad and combined with the reaction product. The iron that remains in the flask and on the filter pad can be destroyed by treatment with water followed by dilute hydrochloric acid.

The filtrate is transferred to a round bottom flask and the solvent and excess iron pentacarbonyl are removed on a rotary evaporator fitted with a solid CO_2/acetone condenser (in the fume hood). The unreacted iron pentacarbonyl freezes onto the condenser surface and the butyl ether collects in the receiver. After removal of all solvent the receiver is replaced with a flask containing bromine water and the iron pentacarbonyl is allowed to thaw; it is oxidized by the bromine and can be discarded safely in the usual way for inorganic residues.

The crude complex thus obtained (140 g, 58% yield) is normally sufficiently pure for hydride abstraction as described in Chapter 5. It may be purified by chromatography on silica gel. Elution with petroleum ether removes isomeric complexes (e.g., the 2-methoxy-5-methylcyclohexa-1,3-dien complex, which is much faster running than **4.11**). The desired complex is eluted with either 25% ethyl acetate in hexane or with ether–petroleum ether (50:50). Removal of solvent on the rotary evaporator affords the complex **4.11** as a yellow viscous oil, which is slowly oxidized by air. Storage under nitrogen in the refrigerator is recommended. The IR spectrum for this complex shows very strong carbonyl stretching absorptions at 2040 and 1965 cm^{-1}. Proton NMR ($CDCl_3$) shows H(2) at δ 5.15 (d, J 4.5 Hz), H(3) at 4.85 (d, J 4.5 Hz), MeO singlet at 3.48 and Me singlet at 1.54. The remaining methylene protons occur as an undefined multiplet at δ 3.0–1.70. The chemical shifts of the vinyl protons are typical of those for "inner" diene hydrogens (i.e., C(2) and C(3)). The "outer" diene protons, H(1) and H(4) are normally found at much higher field, around δ 3.3–2.8.

Preparation of (ergosteryl benzoate)tricarbonyliron (**4.18**)

This procedure illustrates the *in situ* generation of *p*-methoxybenzylidene-acetonetricarbonyliron and its use as an $Fe(CO)_3$ transfer reagent, leading to improved yields of diene-$Fe(CO)_3$ complexes that are difficult to prepare[8]. Ergosteryl benzoate (4.5 g, 9 mmol), *p*-methoxybenzylideneacetone (5.4 g, 17 mmol), and nonacarbonyldiiron (10.9 g, 30 mmol) are stirred in toluene (40 ml) at 55°C under nitrogen for 5 days. The reaction can be followed by TLC (4% petroleum ether–ethyl acetate, four times developed) until all the starting ergosteryl benzoate is consumed. The solution is filtered and evaporated *in vacuo* and the oily residue is triturated with methanol to give a solid which is recrystallized from ethyl acetate to give the complex (4.3 g, 80% yield), m.p. 163–165°C, $[\alpha]_D$ −67.9° (c 1.20, $CHCl_3$); IR: 2040 and 1965 $[Fe(CO)_3]$, 1715 (Ph*CO*). ^{1}H NMR: δ 8.12–7.95 (2H, ArH), 7.60–7.30 (3H, ArH), 5.17 (2H, uncomplexed vinyl), 5.04 (2H, ABq, *J* 4 Hz, complexed vinyl-H), 4.97 (1H, 3-H), 2.8–1.17 (methylenes), 1.07, 0.97, 0.87, 0.77, 0.73 (methyls).

Preparation of troponeiron tricarbonyl **(4.23)**

The procedure is adapted from ref. [71]. Cycloheptatrienone (1.095 g, 10.32 mmol) and nonacarbonyldiiron (8.56 g, 23.5 mmol) are heated for 90 min at 55°C in degassed benzene (20 ml) under nitrogen and with exclusion of light. The solution is cooled to room temperature, concentrated and then chromatographed using a 3 inch column of alumina (neutral, activity 3). The product is eluted as a deep red–orange band using 1:1 methylene chloride/hexane or petroleum ether. Evaporation of solvent affords **4.23** (2.46 g, 96% yield) as a red solid, m.p. 70–71°C; IR (CH_2Cl_2): 2020, 2001, 1950, 1637 cm^{-1}. ^{1}H NMR ($CDCl_3$): δ 6.58 (1H, m), 6.39 (2H, m), 5.05 (1H, m), 3.19 (1H, m), 2.75 (1H, m). The latter two peaks are due to the "outer" diene protons; see the comment on p. 90.

Preparation of tricarbonyl[(1-4-η)-5,6-exo-dihydroxy-7-endo{(t-butyldimethylsilyl) oxy}-cyclohepta-1,3-diene]iron **(4.25)**

This procedure is an example of the use of the $Fe(CO)_3$ group as diene protection during the osmylation of an uncomplexed double bond[9]. The starting material, complex **4.24**, is prepared from tropone-$Fe(CO)_3$ using standard borohydride reduction and OH protection methods (see equation 4.10). (1) Stoichiometric osmylation: to a stirred solution of complex **4.24** (175 mg, 0.48 mmol) in pyridine (1.7 ml) under nitrogen is added osmium tetraoxide (6.8 ml of a 0.0786 M solution in THF). The mixture is stirred at room temperature for 10 h, after which time saturated aqueous sodium bisulfite (10 ml) is added. After stirring for 10 h, the mixture is poured into ethyl acetate (10 ml), filtered through Celite and the pad is washed thoroughly with ethyl acetate. The combined washings are partitioned and the organic layer is washed with cold dilute hydrochloric acid (5%, 10 ml), water (10 ml), aqueous sodium bicarbonate (10 ml), dried ($MgSO_4$) and evaporated to give the complex **4.25** (195 mg, 98% yield) as a pale yellow solid. An analytical sample, m.p. 115–116°C, can be prepared by recrystallization from methylene chloride/hexane. (2) Catalytic osmylation: to a stirred solution of complex **4.24** (7.0 g, 19.3 mmol) in acetone (25 ml) under nitrogen is added tetraethylammonium acetate (1.21 g, 4.8 mmol). After stirring for 5 min at room temp., the mixture is cooled to 0°C and t-butyl hydroperoxide (3.1 ml of 90%, 31 mmol) is added, followed by osmium tetraoxide (12.3 ml of a 0.0786 M solution in THF). The reaction mixture is stirred at 0°C for 1 h, then at room temperature for 1 h, and saturated aqueous sodium bisulfite (25 ml) is added. After stirring for 2 h, the reaction mixture is diluted with 100 ml of ethyl acetate, filtered through a Celite pad, and the filter pad is washed with 50 ml of ethyl acetate. The combined filtrate is washed with water (3 × 100 ml), the organic layer is dried ($MgSO_4$) and evaporated to give the complex **4.25** containing approx. 10% ketol impurity from overoxidation.

*Preparation of tricarbonyl[(1-4-η)-{6-*exo*-hydroxy-7-*endo*-(t-butyldimethylsilyl) oxy}cyclohepta-1,3-diene]iron* **(4.26)**

This preparation gives an example of the use of the $Fe(CO)_3$ group as diene protection during the hydroboration of an uncomplexed double bond[9]. To an ice-cooled solution of complex **4.24** (92 mg, 0.25 mmol) in THF (2 ml) under nitrogen is added 2.5 ml of a 1.0 M solution of borane in THF. The reaction mixture is stirred at 0°C for 4 h, and water (2 ml) followed by sodium hydroxide (200 μl of 15% aqueous solution) and hydrogen peroxide (200 μl of 30% aqueous solution) are added. The reaction mixture is stirred for 3 h and the layers are separated. The aqueous layer is washed with ether (10 ml) and the ether layers are combined, washed with water (3 × 20 ml), dried ($MgSO_4$) and evaporated to give the product (90% yield) which may be purified by preparative layer chromatography, R_f 0.39 (20% EtOAc in hexane).

Ligand exchange: preparation of dicarbonyl(triphenylphosphine)cyclohexa-dieneiron **(4.63)**[28]

A stirred solution of tricarbonylcyclohexadieneiron (**4.62**) (5.0 g) and triphenyl-phosphine (6.0 g) in cyclohexanol (50 ml) is refluxed under nitrogen overnight. The mixture is cooled, light petroleum (b.p. 40–60°C, 100 ml) is added, and the resulting precipitate of $[Fe(CO)_3(PPh_3)_2]$ by-product (1.2–1.7 g) is removed by filtration. The liquors are evaporated, cyclohexanol is removed by distillation at 0.1 mm Hg, and the residue is recrystallized from hexane to give complex **4.63** as a yellow crystalline solid (7.3 g, 71% yield), m.p. 120–121°C. The IR spectrum of complex **4.63** shows carbonyl stretching bands at 1970 and 1910 cm^{-1}, which are at lower frequency than those for the tricarbonyl complex **4.62** (2060 and 1960 cm^{-1}), consistent with the poorer π-acceptor ability of the triphenylphosphine versus CO ligand.

Friedel–Crafts acetylation of complex **4.63**[28]

A solution of the Perrier complex in methylene chloride is prepared by gentle warming of acetyl chloride (1.2 ml) and aluminum chloride (2.94 g) followed by addition of anhydrous methylene chloride (10 ml) to the cooled melt. To a stirred solution of complex **4.63** (0.50 g, 1.11 mmol) in dry methylene chloride (10 ml) under argon at −78°C is added 1 ml of the Perrier complex solution. At half hourly intervals, two further 1ml portions of the Perrier complex are added. After the second addition, stirring is continued for 10 min, the reaction mixture is diluted with dry ether at −78°C, and is poured onto ice. The organic layer is separated, washed with water (4 × 15 ml), brine (20 ml), dried ($MgSO_4$) and evaporated to give an oily product which is recrystallized from methylene chloride/pentane to afford complex **4.64** (0.53 g, 96% yield), m.p.128–129°C.

Diene/olefin coupling reaction: cyclization of **4.73** *to give* **4.74**[35]

The amide is prepared by using standard methodology from the corresponding carboxylic acid, which is converted into the acid chloride (oxalyl chloride, pyridine) and this is treated with the amine *in situ*. The allyl amide **4.73** (200 mg, 0.51 mmol) is dissolved in di-*n*-butyl ether (10 ml) in a clean, flame-dried single neck flask. A magnetic stir bar is placed in the flask, the solution is purged with carbon monoxide (*in a fume hood!*) for 1 min, and a reflux condenser is attached while still purging with CO. To the top of the condenser is attached a balloon of carbon monoxide (do not over pressurize), or a carbon monoxide bubbler, and the stirred solution is then heated at gentle reflux for 6.5 h. The solution is cooled to room temperature, diluted with diethyl ether and filtered through Celite. Concentration of the filtrate, followed by preparative layer chromatography (silica gel, 20% ethyl acetate in hexane), affords the spirolactam **4.74** (180–184 mg, 90–92% yield), m.p. 143–144°C. The IR spectrum shows the typical $Fe(CO)_3$ bands at 2020 and 1980 cm^{-1}, as well as the lactam carbonyl stretch at 1700 cm^{-1}.

Nucleophile addition/cyclocarbonylation of butadienetricarbonyliron: conversion of **4.1** *into* **4.105**

The procedure is adapted from ref. [42]. To a 0.2 M stirred solution of lithium diisopropylamide (1.4 mol equiv.) in THF at −78°C under argon is added via syringe isobutyronitrile (1.5 mol equiv.) over a period of 30 s. The carbanion solution so generated is used in the nucleophile addition reaction immediately, as follows. A slow stream of carbon monoxide is passed over a stirred solution of the anion (2.3 mmol) in THF (10 ml) cooled to −78°C (*fume hood!*). Use two syringe needles inserted through a rubber septum, one for entry of CO and one for exit; do not over-pressurize the reaction vessel, which is a one-neck flask equipped with a magnetic stirrer. A solution of butadiene-$Fe(CO)_3$ (**4.1**) (2.3 mmol) in THF (0.5 ml) is added dropwise over a period of 1 min, then the gas exit needle is withdrawn and the CO pressure is increased to *ca* 8–15 psi above 1 atm. The septum should be secured using copper wire so as to maintain the modest positive pressure. The mixture is stirred for 2 h at −78°C, allowed to warm to room temperature, then cooled back to −78°C. Trifluoroacetic acid (1.0 ml, 13 mmol) is added and the solution is stirred for 5 min at −78°C, after which time it is poured into aqueous acetone. Ceric ammonium nitrate is added in small portions until there is no further evolution of carbon monoxide and the mixture is partitioned between water and ether. The ether layer is removed, washed with water, dried and evaporated in the usual way and the product is purified by short-path distillation, using a bath temperature of 80–110°C at 0.8 Torr. The product **4.105** is isolated as a colorless oil in 78% yield.

Cyclocarbonylation of 1,6-diynes: preparation of complex **4.115**[52,53]

A solution of 1,7-bis(trimethylsilyl)-1,6-heptadiyne (0.482 g, 2.04 mmol) in toluene (10 ml) is injected into a 60-ml quartz Griffen–Worden pressure vessel (Kontes Glassware), followed by pentacarbonyliron (5.0 ml, 18 mmol). The solution is degassed by the freeze–pump–thaw method. After warming to room temperature, the vessel is charged with carbon monoxide (100 psi), and is placed in an oil bath at 125–130°C for 24 h. After cooling and releasing the pressure, the reaction mixture is diluted with methylene chloride and filtered through Celite. The solvent is removed on the rotary evaporator, using a solid CO_2/acetone trap to freeze any small amounts of residual pentacarbonyliron (see p. 90). The solid residue is dried *in vacuo* and the product is purified by flash chromatography, R_f = 0.32 (10% EtOAc in hexane), to give the complex **4.115** as a yellow crystalline solid, m.p. 147–148°C; IR (CCl_4) 2064, 2006, 1984, and 1624 cm^{-1}.

Preparation of tricarbonyl(benzylideneacetone)iron (**4.126**)[62]

Benzylideneacetone (1.04 g) and nonacarbonyldiiron (2.60 g) are stirred in toluene under nitrogen at 60°C for 4 h. The solution is cooled to room temperature, filtered through silica gel and the solvent is removed on the rotary evaporator. The residue is chromatographed on silica gel, eluting with toluene, to afford the complex **4.126** as red crystals (0.61 g, 32% yield), m.p. 88–90°C; IR 2065, 2005, 1985 cm^{-1}.

References

1. H. Reihlen, A. Gruhl, G. von Hessling, and O. Pfrengle, *Liebigs. Ann. Chem.* **482**, 161 (1930).
2. A.J. Birch and G. Subba Rao, *Adv. Org. Chem.* **8**, 1 (1972).
3. J.M. Hook and L.N. Mander, *Nat. Prod. Reports* **3**, 35 (1986).
4. A.J. Birch, P.E. Cross, J. Lewis, D.A.White, and S.B. Wild, *J. Chem. Soc. (A)* 332 (1968).
5. A.J. Birch and K.P. Dastur, *J. Chem. Soc. Perkin Trans.* **1**, 1650 (1973).
6. A.J. Pearson and C.W. Ong, *J. Am. Chem. Soc.* **103**, 6686 (1981).
7. G. Evans, B.F.G. Johnson and J. Lewis, *J. Organomet. Chem.* **102**, 507 (1975).
8. D.H.R. Barton, A.A.L. Gunatilaka, T. Nakanishi, H. Patin, D.A. Widdowson and B.R. Worth, *J. Chem. Soc. Perkin Trans.* **1**, 821 (1976).
9. A.J. Pearson and K. Srinivasan, *J. Org. Chem.* **57**, 3965 (1992).
10. A.J. Pearson and Y.S. Chen, *J. Org. Chem.* **57**, 1939 (1986).
11. K.B. Sharpless and R.C. Michaelson, *J. Am. Chem. Soc.* **95**, 6136 (1973).
12. R.G. Carlson and N.S. Behn, *J. Org. Chem.* **32**, 1363 (1967).
13. M. Franck-Neumann and D. Martina, *Tetrahedron Lett.* 1759 (1975).
14. Y. Shvo and E. Hazum, J. *Chem. Soc. Chem. Commun.* 336 (1974).
15. B.F.G. Johnson, J. Lewis, P. McArdle and G.L.P. Randall, *J. Chem. Soc. Dalton Trans.* 456 (1972).
16. J.H. Rigby and C.O. Ogbu, *Tetrahedron Lett.* **31**, 3385 (1990).
17. M. Franck-Neumann, F. Brion and D. Martina, *Tetrahedron Lett.* 5033 (1978).
18. D. Lloyd, *Carbocyclic Non-benzenoid Aromatics*, pp. 135,144. Elsevier, Amsterdam, 1966,
19. R. Grée, *Synthesis* 341 (1989).
20. A.J. Birch and A.J. Pearson, *J. Chem. Soc. Chem. Commun.* 6601 (1976).
21. E.O. Greaves, G.R. Knox and P.L. Pauson, *J. Chem. Soc. Chem. Commun.* 1124 (1969).
22. E.O. Greaves, G.R. Knox, P.L. Pauson, S. Toma, G.A. Sim and D.I. Woodhouse, *J. Chem. Soc.*

Chem. Commun. 257 (1974).

23. R.E. Graf and C.P. Lillya, *J. Organomet. Chem.* **166**, 53 (1979).
24. R.E. Graf and C.P. Lillya, *J. Organomet. Chem.* **122**, 377 (1976).
25. A.J. Pearson, *Transition Met. Chem.* **6**, 67 (1981).
26. N.S. Nametkin, A.I. Nekhaev, V.D. Tyurin and S.P. Gubin, *Izv. Akad. Nauk, SSSR Ser. Khim.* 676 (1975).
27. B.F.G. Johnson, J. Lewis and D. Parker, *J. Organomet. Chem.* **141**, 319 (1977).
28. A.J. Birch, S.Y. Hsu, A.J. Pearson and W.D. Raverty, *J. Organomet. Chem.* **260**, C59 (1984).
29. A. Bond and M. Green, *J. Chem. Soc.* 12 (1971).
30. A. Bond, M. Green and S.F.W. Loweie, *J. Chem. Soc. Chem. Commun.* 1230 (1971).
31. A. Bond, M. Green and S.H. Taylor, *J. Chem. Soc. Chem. Commun.* 112 (1973).
32. A. Bond and M. Green, *J. Chem. Soc. Dalton Trans.* 763 (1974).
33. A. Bond, B. Lewis and M. Green, *J. Chem. Soc. Dalton Trans.* 1109 (1975).
34. M. Green, B. Lewis, J.J. Daly and F. Sanz, *J. Chem. Soc. Dalton Trans.* 1118 (1975).
35. A.J. Pearson and M.W. Zettler, *J. Am. Chem. Soc.* **111**, 3908 (1989).
36. J.M. Takacs and L.G. Anderson, *J. Am. Chem. Soc.* **109**, 2200 (1987).
37. W.R. Roush and J.C. Park, *Tetrahedron Lett.* **31**, 4707 (1990).
38. S.G. Davies, M.L.H. Green and D.M.P. Mingos, *Tetrahedron* **34**, 3047 (1978).
39. A.J. Pearson, in *Advances Metal-Organic Chemistry* (ed. L.S. Liebeskind), Vol. I, Chapter 1, JAI Press, Greenwich, CT, 1989.
40. M.F. Semmelhack and J.W. Herndon, *Organometallics* **2**, 363 (1983).
41. M.F. Semmelhack, *Pure Appl. Chem.* **53**, 2379 (1981).
42. M.F. Semmelhack, J.W. Herndon and J.K. Liu, *Organometallics* **2**, 1885 (1983).
43. H.C. Longuet-Higgins and L.E. Orgel, *J. Chem. Soc.* 1969 (1956).
44. G.F. Emerson, L. Watts and R. Pettit, *J. Am. Chem. Soc.* **87**, 131 (1965).
45. R. Pettit, J.C. Barborak and L. Watts, *J. Am. Chem. Soc.* **88**, 1328 (1966).
46. P.E. Eaton and T.W. Cole, Jr, *J. Am. Chem. Soc.* **86**, 962 (1964).
47. S. Masamune, C.G. Chin and H.W. Cuts, *J. Chem. Soc. Chem. Commun.* 880 (1966).
48. E. Weiss, R.G. Merenyi and W. Hubel, *Chem. Ind. (London)* 407 (1960).
49. E. Weiss, R.G. Merenyi and W. Hubel, *Chem. Ber.* **95**, 1170 (1962).
50. R.S. Dickson, C. Mok and G. Connor, *Aust. J. Chem.* **30**, 2143 (1977).
51. E.R.F. Gesing, J.P. Tane and K.P.C. Vollhardt, *Angew. Chem. Int. Ed. Engl.* **19**, 1023 (1980).
52. A.J. Pearson and R.A. Dubbert, *J. Chem. Soc. Chem. Commun.* 202 (1991).
53. A.J. Pearson, R.A. Dubbert and R.J. Shively, Jr, *Organometallics* **11**, 4096 (1992).
54. Pentalenolactone: B. Koe, B.A. Sobin and W.D. Celmer, *Antibiot. Annu.* 672 (1957).
55. Coriolin: P.F. Schuda, H.L. Ammon, M.R. Heimann and S. Battacharjee, *J. Org. Chem.* **47**, 3434 (1982).
56. Subergorgic acid: A. Groweiss, W. Fenical, C. He, J. Clardy, Z. Wu, Z. Yiao and K. Long, *Tetrahedron Lett.* **26**, 2379 (1985).
57. Pentalene: H. Seto and H. Yonehara, *J. Antibiot.* **33**, 92 (1980).
58. Cyclopentanoid and polyquinane synthesis: L.A. Paquette, *Top. Curr. Chem.* **119**, 1 (1984).
59. Cyclopentanoid and polyquinane synthesis: *Tetrahedron* **37**, 4359 (1981).
60. A.J. Birch, K.B. Chamberlain, M.A. Haas and D.J. Thompson, *J. Chem. Soc. Perkin Trans.* **1**, 1882 (1973).
61. G.R. Stephenson and D.A. Owen, *Tetrahedron Lett.* **32**, 1291 (1991).
62. A.M. Brodie, B.F.G. Johnson, P.L. Josty and J. Lewis, *J. Chem. Soc. Dalton Trans.* 2031 (1972).
63. S.E. Thomas, T.N. Danks and D. Rakshit, *Phil. Trans. R. Soc. Lond. A*, **326**, 611 (1988).
64. T.N. Danks, D. Rakshit and S.E. Thomas, *J. Chem. Soc. Perkin Trans.* **1**, 2091 (1988).
65. N.W. Alcock, T.N. Danks, C.J. Richards and S.E. Thomas, *J. Chem. Soc. Chem. Commun.* 21 (1989).
66. T.N. Danks and S.E. Thomas, *J. Chem. Soc. Perkin Trans.* **1**, 761 (1990).
67. C.J. Richards and S.E. Thomas, *J. Chem. Soc. Chem. Commun.* 307 (1990).

68. L. Hill, C.J. Richards and S.E. Thomas, *J. Chem. Soc. Chem. Commun.* 1085 (1990).
69. K.G. Morris and S.E. Thomas, *J. Chem. Soc. Perkin Trans.* **1**, 97 (1991).
70. A. Ibbotson, A.C. Redutodos Reis, S.P. Saberi, A.M.Z. Slawin, S.E. Thomas, G.J. Tustin and D.J. Williams, *J. Chem. Soc. Perkin Trans.* **1**, 1251 (1992).
71. M. Rosenblum and J.C. Watkins, *J. Am. Chem. Soc.* **112**, 6316 (1990).

–5–

Dienyl Complexes of Iron

This chapter is divided into two major sections, the first dealing with the chemistry and synthetic applications of open cationic dienyliron complexes, and the second dealing with ferrocene and its derivatives, owing to the different reactivity patterns and, therefore, applications of the two types of complex. The open dienyl complexes are powerful electrophiles and have found many applications in natural products synthesis, whereas ferrocene is nucleophilic and its derivatives have been applied in materials science and as asymmetric ligands in a number of catalytic processes. Together, these molecules probably constitute the largest group of organoiron systems and continue to be actively investigated in organic synthesis. They are also some of the most stable and easy to handle organometallic complexes, since they do not require the use of specialized apparatus such as glove boxes and Schlenk lines for their manipulation. Standard organic synthesis techniques are typically used for running reactions and isolating the products. Moreover, the open dienyl complexes are easily prepared on large scale using inexpensive reagents, and usually require no purification other than precipitation from the reaction mixture and filtration. These cationic complexes may in some ways be likened to very reactive, highly functionalized, alkyl halides, that show exceptional selectivity during their reactions with nucleophiles.

5.1 OPEN PENTADIENYLIRON COMPLEXES

There are many examples of these compounds, ranging from acyclic pentadienyl systems to cyclic ligands such as cyclohexadienyl and cycloheptadienyl. By far the most studied are the cyclohexadienyl complexes, although there has been increased interest in the acyclic and seven-membered ring counterparts over the past ten years. Consequently, the six-membered ring systems will be considered first.

5.1.1 Cyclohexadienyl-$Fe(CO)_3$ complexes: preparation

The parent complex (**5.2**) was the first of this class to be described[1]. Treatment of cyclohexadiene-$Fe(CO)_3$ (**5.1**) with triphenylmethyl (trityl) tetrafluoroborate in methylene chloride solution at room temperature for *ca* 45 min leads to virtually quantitative conversion into **5.2**. This is a most interesting reaction. On admixture of methylene chloride solutions of the two reactants (both yellow) a dark brown coloration is produced, possibly due to the formation of a charge transfer complex. Gradually the color lightens, becoming golden yellow, and the tetrafluoroborate **5.2** precipitates out. This complex can be removed by filtration. Alternatively, precipitation can be completed by pouring the reaction mixture into diethyl ether which has not been dried. The product is filtered off and washed thoroughly with "wet" ether, the latter serving a dual purpose in removing ether-soluble triphenylmethane and in hydrolyzing any excess trityl cation to ether-soluble triphenylmethanol. The dienyl complex so obtained is usually analytically pure after drying in air or *in vacuo*. It is quite stable and can be stored in a standard reagent jar for long periods.

$Fe(CO)_3$ **5.1** $\xrightarrow{Ph_3CBF_4}$ $Fe(CO)_3$ BF_4^- **5.2** (5.1)

Triphenylmethyl tetrafluoroborate, the hydride abstracting reagent, is readily prepared by treating triphenylmethanol dissolved in acetic anhydride with 48% aqueous tetrafluoroboric acid. Since the accompanying acid-catalyzed reaction of water with acetic anhydride is exothermic, the addition is carried out dropwise to the stirred triphenylmethanol solution cooled in an ice bath. The trityl tetrafluoroborate is precipitated with dry ether, filtered off and washed with dry ether. It is very moisture sensitive and should be stored in a tightly sealed jar in the refrigerator. If low yields are encountered during the preparation of the trityl tetrafluoroborate, it is usually an indication that the triphenylmethanol is of poor quality, and this is easily remedied by recrystallization from hot methanol. Also, some of the more highly substituted cyclohexadienyl complexes do not give easily crystallizable tetrafluoroborate derivatives, these sometimes being thrown down as gums upon addition of ether to the methylene chloride reaction solution (or vice versa). This is usually taken care of by conversion of the water-soluble tetrafluoroborates into water-insoluble hexafluorophosphates by treatment with aqueous ammonium hexafluorophosphate. Even better in such instances is to prepare the complex by using trityl hexafluorophosphate directly, which can be prepared in analogous fashion to the tetrafluoroborate by substituting HPF_6 for HBF_4.

A number of functionally substituted cyclohexadienyliron complexes have been prepared in this manner. The hydride abstraction reaction shows quite remarkable selectivity, according to the nature of the substituent on the diene, and usually this is the opposite of what would have been expected intuitively, based on what is understood about the corresponding uncomplexed dienyl cations. In order to illustrate this point, let us first consider a well-known organic reaction, electrophilic aromatic substitution. As is known, electron-donating substituents on the benzene ring favor *ortho*/*para* attack, whereas electron-withdrawing substituents favor *meta* attack. This can be explained on the basis of the effect of the substituent on the intermediate dienyl cations. Electron-donating groups stabilize the dienyls resulting from *ortho* and *para* attack, whereas these are destabilized by electron-withdrawing groups (EWG). In the latter case, therefore, a dienyl having the EWG at C-2 is favored. The situation is exactly the opposite for dienyl-$Fe(CO)_3$ complexes, although the hydride abstraction reaction is also sensitive to steric effects and may in practice be somewhat more complicated than is implied here. The overall picture is given in Fig. 5.1.

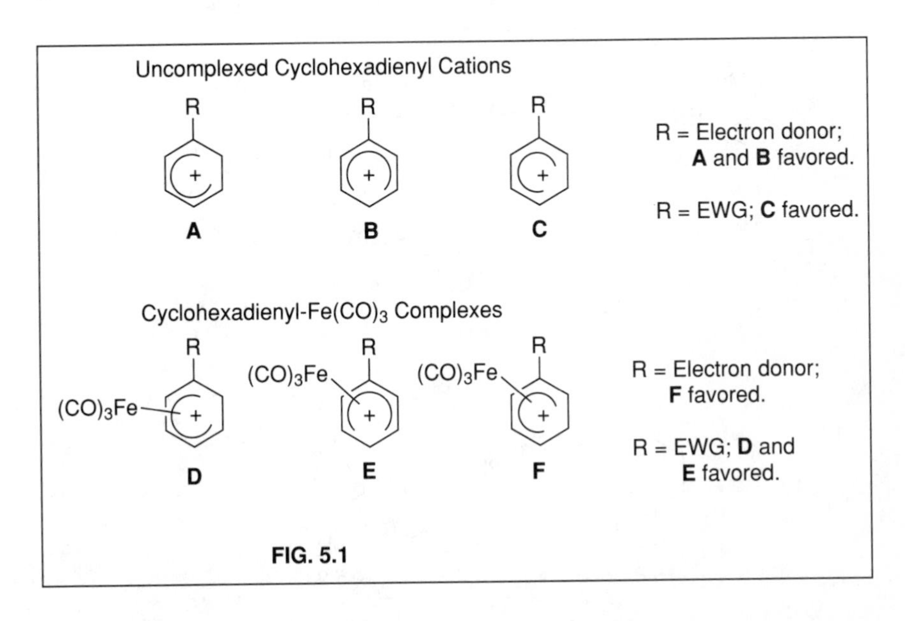

FIG. 5.1

The explanation of these directing effects lies within the nature of the orbital interactions between the dienyl ligand and the $Fe(CO)_3$ group[2]. The well-known synergistic interaction between metal and ligand is characterized by overlap between the highest occupied molecular orbital (HOMO) of the dienyl and an empty metal d orbital (or hybrid), and an overlap between dienyl lowest unoccupied molecular orbital (LUMO) and a filled metal orbital. This interaction will be strongest when the interacting orbitals are close in energy. It turns out that more stable uncomplexed dienyls have lower energy HOMO (as expected intuitively) and higher energy LUMO than the less stable dienyls. Consequently, the latter give a better energy match with the iron d orbitals and, as a result, a stronger bonding interaction. This thermodynamic favorability is reflected in the transition state for hydride abstraction which, because of the developing positive charge, must show the same stabilizing effects by substituents on the diene. Therefore, the activation energy is lower for the formation of the dienyl complexes indicated in Fig. 5.1. In practice, the results of hydride abstraction are as shown in equations 5.2–5.7[3–6]. Preparations of some of the diene complexes that are used here will be outlined in the later sections which deal with actual synthetic applications of the dienyl complexes.

OMe Fe(CO)3 — Ph_3CBF_4 → (CO)3Fe OMe + OMe Fe(CO)3 BF_4^- BF_4^- (5.2)

5.3 **5.4** (94%) **5.5** (6%)

OMe Fe(CO)3 — Ph_3CBF_4 → (CO)3Fe OMe + OMe Fe(CO)3 BF_4^- BF_4^- (5.3)

5.6 **5.4** (20%) **5.7** (80%)

OMe Fe(CO)3 Me — Ph_3CBF_4 → OMe Fe(CO)3 BF_4^- Me (5.4)

5.8 **5.9** (95%)

CO_2Me Fe(CO)3 — Ph_3CBF_4 → CO_2Me Fe(CO)3 BF_4^- (5.5)

5.10 **5.11** (95%)

R Fe(CO)3 — Ph_3CBF_4 → R Fe(CO)3 BF_4^- + R Fe(CO)3 BF_4^- (5.6)

5.12 **5.13** **5.14**

(a) R = CO_2Me (a) 80% (a) 20%

(b) R = $SiMe_3$ (b) 100% (b) 0%

$$\mathbf{5.5} + \mathbf{5.16} \xrightarrow{Ph_3CBF_4} \mathbf{5.17} + \mathbf{5.18} + \mathbf{5.19} \quad (5.7)$$

The results are largely what is expected from the preceding discussion. Complex **5.6** is somewhat anomalous, giving **5.7** as the major product. This is most likely due to steric effects since, in order to form **5.4**, the bulky trityl cation must approach a hydrogen closer to the methoxy substituent. The true extent of the electronic driving force for the hydride abstraction reaction is best judged from the results obtained with the sterically unbiased diene complexes **5.3** and **5.12**. Indeed, when steric hindrance at both methylenes is approximately balanced, as in complex **5.8**, the major product (**5.9**) is the electronically favored one. As we shall see later, this is extremely fortuitous from the standpoint of organic synthesis.

The effect of a methyl substituent on the hydride abstraction is rather difficult to determine, since the reaction of dihydrotoluene with $Fe(CO)_5$ gives an approximately 2:1 mixture of complexes **5.15** and **5.16**. Treatment of this mixture with trityl tetrafluoroborate gives an approximately equimolar mixture of the three complexes shown in equation 5.7. For synthetic applications this is most undesirable, but with a little ingenuity the problem can be overcome. It has been found[7] that treatment of diene complexes, derived from *ortho-*, *meta-* or *para-* cresol, with concentrated sulfuric acid at 0–5°C, followed by thorough washing with ether and anion exchange

$$\text{2-methylanisole} \xrightarrow[2)\ Fe(CO)_5,\ \Delta]{1)\ Li,\ NH_3} \mathbf{5.20} + \mathbf{5.21} + \mathbf{5.22} \xrightarrow{1)\ c.\ H_2SO_4\ \ 2)\ NH_4PF_6} \mathbf{5.18}\ (PF_6^-) \quad (5.8)$$

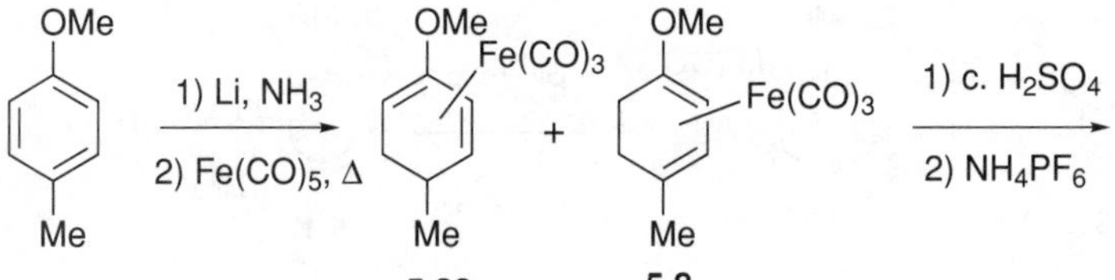

OMe
Me
1) Li, NH_3
2) $Fe(CO)_5$, Δ
OMe $Fe(CO)_3$ Me **5.23** + OMe $Fe(CO)_3$ Me **5.8**
1) c. H_2SO_4
2) NH_4PF_6
5.18 (5.9)

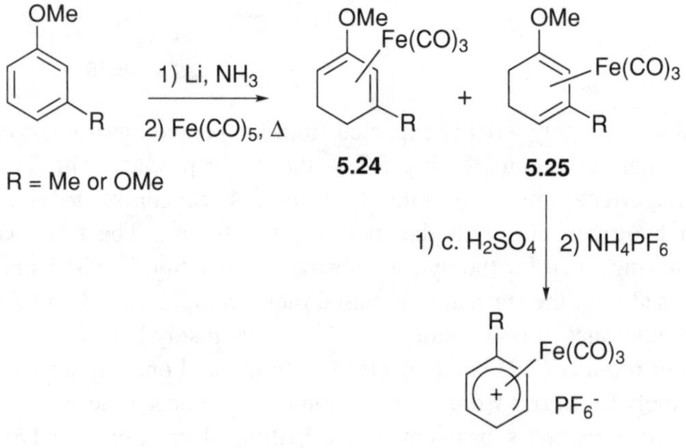

OMe R
R = Me or OMe
1) Li, NH_3
2) $Fe(CO)_5$, Δ
OMe $Fe(CO)_3$ R **5.24** + OMe $Fe(CO)_3$ R **5.25**
1) c. H_2SO_4 2) NH_4PF_6
R $Fe(CO)_3$ + PF_6^-
5.5 R = OMe
5.19 R = Me
(5.10)

using aqueous ammonium hexafluorophosphate, give regiochemically defined methyl-substituted dienyl complexes. Even mixtures of diene complexes give a single dienyl product, and the results are summarized in equations 5.8–5.10. This method also allows access to complex **5.5** which is otherwise difficult to obtain (see equation 5.10).

Whereas the dienyl complexes **5.18** and **5.19** can be readily obtained using this

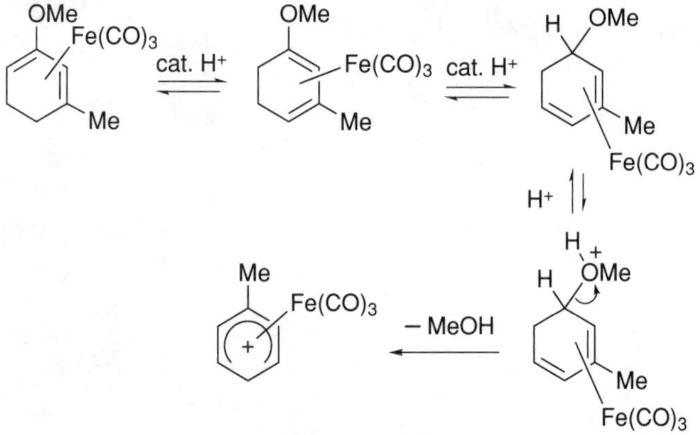

OMe $Fe(CO)_3$ Me
cat. H^+
OMe $Fe(CO)_3$ Me
cat. H^+
H OMe Me $Fe(CO)_3$
H^+
H + H OMe Me $Fe(CO)_3$
– MeOH
Me $Fe(CO)_3$ +

SCHEME 5.1

method, the 1-methylcyclohexadienyl complex **5.17** is inaccessible from a substituted anisole. This is implicit in the mechanism of the demethoxylation process, which is outlined in Scheme 5.1. Acid-catalyzed rearrangement of the diene complex occurs until the methoxy substituent ends up on an sp^3 carbon, at which point protonation on oxygen, followed by rapid irreversible loss of methanol, gives the substituted dienyl complex.

Birch and Williamson[5] discovered that treatment of the hydroxymethyl-substituted complex **5.26** with tetrafluoroboric acid in propionic anhydride gave exclusively the dienyl complex **5.17** in reasonable yield (55%). The mechanism of this conversion is probably as outlined in Scheme 5.2, involving dehydration with concomitant hydride shift, driven by the stability of the resulting dienyliron system. This type of 1,2- shift has also been observed during attempts to acylate complex **5.27** (see Chapter 4), which results only in the formation of the dienyl complex **5.28**. (equation 5.11)[8].

The dehydration processes outlined above have also been employed for the preparation of optically pure dienyl-$Fe(CO)_3$ complexes. The observant reader will have noticed that cyclohexadienyl complexes having substituents at C-1 or C-2 are chiral molecules and, in principle, can be used as precursors in asymmetric synthesis provided they can be generated in non-racemic form. Methods have been

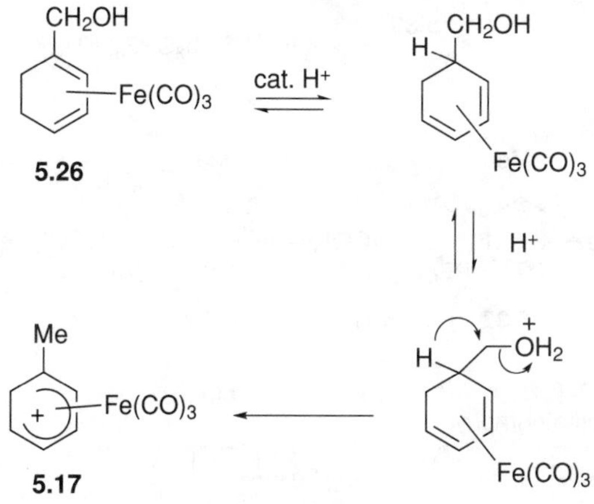

SCHEME 5.2

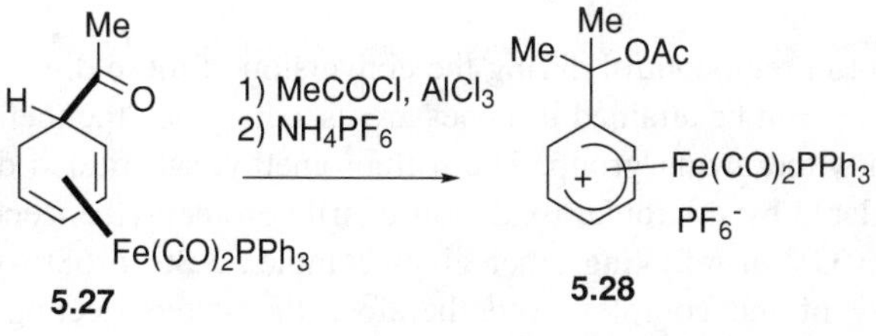

(5.11)

described[9–12] for preparation of optically pure complexes of cyclohexadiene diols which capitalize on two observations: (1) microbial hydroxylation of substituted benzenes gives optically pure cyclohexadiene diols[13]; (2) hydroxyl and alkyl ether groups show a very pronounced stereodirecting effect during reaction of the diene diols and their ethers with $Fe_2(CO)_9$. Thus, the diol **5.29** is converted into the dimethyl ether **5.30**, and treatment of this with $Fe_2(CO)_9$ in refluxing ether gives complex **5.31** in 53% yield as a single diastereomer. Demethoxylation of **5.31** is effected with trityl cation, but this gives a mixture of complexes **5.32**, **5.33** and **5.34**, the latter being a result of competing hydride abstraction followed by hydrolysis of the resulting 1-methoxycyclohexadienyl complex during work-up. Although the major product is **5.32**, the formation of a mixture is a slight disadvantage (see later). The mixture of complexes **5.32** and **5.33** was converted in three steps (20% overall yield) into optically pure complex (+)-**5.18**, having the absolute stereochemistry shown in the structure (Scheme 5.3).

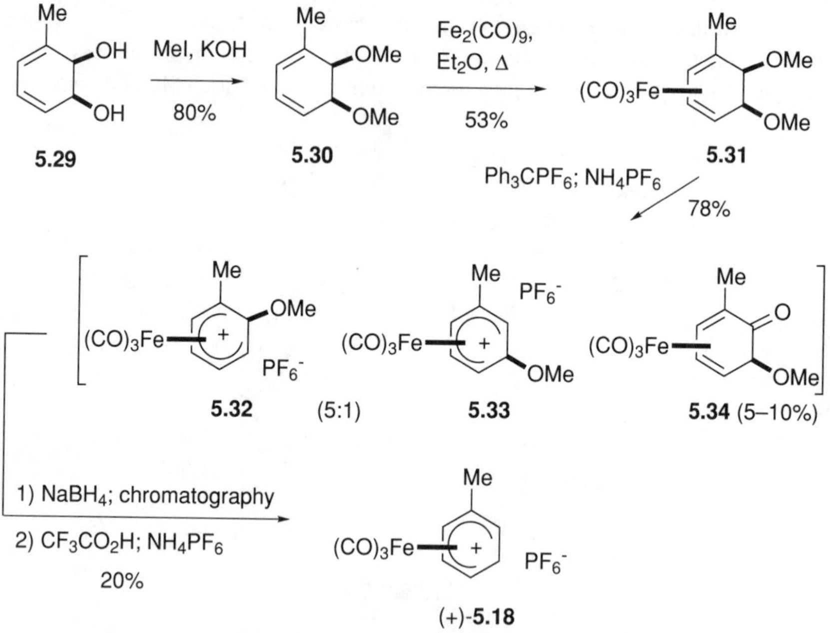

SCHEME 5.3

Improvement in regiocontrol during the conversion of these diol derivatives into dienyl complexes can be attained if dienes are used in which the methyl is replaced by an electron-withdrawing group. The trifluoromethyl-substituted derivative **5.35** is readily available by microbial oxidation of trifluoromethylbenzene. Reaction of **5.35** with $Fe_2(CO)_9$ in refluxing ether gives complex **5.36** in 68% yield[14]. The stereochemistry of the complex, and therefore the stereodirecting effect of the hydroxyls, was confirmed by X-ray crystallography. Treatment of **5.36** with hexafluorophosphoric acid in methylene chloride, in the presence of acetic anhydride, leads to dehydration and *in situ* acetylation of the remaining hydroxyl to give the single dienyl complex **5.37** in 72% yield. On the other hand, reaction of **5.36** with trityl hexafluorophosphate gives exclusively the dienone complex **5.38**, the product of regiospecific hydride abstraction.

CF_3, OH, OH — **5.35** → ($Fe_2(CO)_9$, Et_2O, Δ; 68%) → $(CO)_3Fe$, CF_3, OH, OH — **5.36** → (HPF_6, Ac_2O, CH_2Cl_2; 72%) → CF_3, OAc, +, PF_6^-, $(CO)_3Fe$ — **5.37** (5.12)

5.36 → (Ph_3CPF_6; 85%) → $(CO)_3Fe$, CF_3, OH, O — **5.38** (5.13)

The hydride abstraction reaction, because of the steric requirements of the trityl cation, also does not work with cyclohexadiene complexes that have a substituent at C-5, *anti* to the metal. Such complexes are readily obtained by nucleophile addition to unsubstituted cyclohexadienyl complexes, as discussed in the next section. This limitation precludes the use of trityl in a sequential nucleophile addition/reactivation/nucleophile addition as shown schematically in equation 5.14. (It should be noted that this sequence can be performed on cycloheptadiene complexes, see p. 145.) However, an alternate pathway has been devised for conversion of certain 5-*exo*-substituted diene complexes into dienyls via an oxidative cyclization[15].

$Fe(CO)_3$, +, BF_4^- → $Fe(CO)_3$, R^1 → (Ph_3CBF_4, blocked) → $Fe(CO)_3$, +, R^1, BF_4^- → $Fe(CO)_3$, R^1, R^2 (5.14)

Treatment of the keto ester adduct **5.39** with manganese dioxide in refluxing benzene gives the product **5.40** resulting from oxidative cyclization of the enol form of the side chain. The ether in this complex is similar to that in complex **5.31** and it is expected that the cyclization products can be converted into dienyl complexes by treatment with acid. Modification of the oxidative cyclization, using an alcohol as the nucleophilic side chain[16,17], allows access to the desired complexes, two examples being given in equations (5.16) and (5.17). In this way highly functionalized dienyl complexes are readily available.

MnO_2, PhH, Δ

5.39 → **5.40** (5.15)

$Tl(CF_3CO_2)_3$ 84%; HBF_4, Ac_2O 92%

5.41 → **5.42** → **5.43** BF_4^- (5.16)

MnO_2, PhH, Δ 75–80%; HPF_6, Ac_2O 80–85%

5.44 → **5.45** → **5.46** PF_6^- (5.17)

5.1.2 Cyclohexadienyliron complexes: reactions with nucleophiles

5.1.2.1 General considerations

With very few exceptions the reaction between a cyclohexadienyliron complex and a nucleophile is highly regioselective, attack occurring at a dienyl terminus, and completely stereoselective, *anti* to the metal[18,19]. The resulting diene complex can be demetallated using one of a number of oxidizing agents: trimethylamine-*N*-oxide[20]; ceric ammonium nitrate[21]; copper(II) chloride[22]; ferric chloride[23]; pyridinium chlorochromate[24–26]. All but the first of these reagents generate acidic reaction conditions and are problematic when acid-sensitive functionality is present in the molecule. By far the most useful reagent in this respect is trimethylamine-*N*-oxide. The overall sequence of nucleophile addition/demetallation leads to very useful methodology for the synthesis of substituted cyclohexadienes, and any synthetic plan calling for such an intermediate can be accommodated using this chemistry. In general terms, the sequence is as shown in equation 5.18.

$Fe(CO)_3$ BF_4^- **5.2** $\xrightarrow{X^-}$ $Fe(CO)_3$ **5.47** $\xrightarrow[\text{or Ce(IV), or Fe(III), or PCC}]{Me_3NO, \text{ or } CuCl_2}$ **5.48** (5.18)

One of the most important considerations in organic synthesis is functional group selectivity. If $Fe(CO)_3$ is regarded as a functional group, it becomes important to establish whether interconversions of other functionality on the complex can be carried out without destruction of the $Fe(CO)_3$ group, and whether the latter can be removed without detriment to sensitive organic functionality. It has been shown in Chapter 4 that the diene-$Fe(CO)_3$ moiety is indeed fairly robust, and is able to withstand reactions conditions such as osmylation, hydroboration and epoxidation of alkene double bonds, even though these involve the use of oxidizing conditions. The present chapter will show that fairly complex functionality can be built into the molecule, in the presence of the diene-$Fe(CO)_3$ group, and demetallation can be effected selectively at a chosen point in the synthetic strategy.

As we have seen, a wide range of substituted cyclohexadienyliron complexes are accessible, in which electron-donating or -withdrawing substituents are attached to the dienyl. Of these, perhaps the most profound regiodirecting effect is associated with a methoxy group at C-2, as in complexes **5.4** and **5.9**. Complexes of structure **5.7** also exhibit potentially useful regiocontrol during nucleophile addition. Nucleophile additions to the 3-methoxydienyl complex **5.5** and to the 1- and 3-methoxycarbonyl dienyl complexes **5.11** and **5.13** are also useful. Generic examples for each of these complexes are given in equations 5.19–5.23. The directing effect observed with the 2-methoxydienyl complexes is undoubtedly due to electronic deactivation of the C-1 position, as a result of electron donation from the methoxyl[2], and extends to complexes having a substituent at the more reactive C-5 terminus (e.g., **5.9**). As will be shown later, this is most useful from the standpoint of organic synthesis. For complexes analogous to **5.9** where R = CH_2R^1 (i.e. longer alkyl side chain) the increased steric hindrance leads to a higher proportion of nucleophile addition at the unsubstituted terminus. In these cases, switching to the *iso*propoxy (instead of MeO) derivatives secures addition to the substituted terminus (see later).

OMe, Fe(CO)3, +, BF4⁻, R — X⁻ → OMe, Fe(CO)3, R, X (5.19)

5.4 R = H
5.9 R = Me
5.49

OMe, + Fe(CO)3, BF4⁻ — X⁻ → X, OMe, Fe(CO)3 (5.20)

OMe, Fe(CO)3, +, BF4⁻ — X⁻ → OMe, Fe(CO)3, X (5.21)

5.5 **5.51**

CO2Me, + Fe(CO)3, BF4⁻ — X⁻ → CO2Me, Fe(CO)3, X (5.22)

5.11 **5.52**

CO2Me, Fe(CO)3, +, BF4⁻ — X⁻ → CO2Me, Fe(CO)3, X (5.23)

5.13 **5.53**

The method chosen for decomplexation of the methoxy-substituted derivatives **5.49** and **5.51** can lead to quite different products, owing to the different reaction conditions and, therefore, can lead to different avenues for potential synthetic application. For example, treatment of **5.49** (R=H) with trimethylamine-*N*-oxide produces cyclohexadienes of structure **5.54**. The acid-sensitive vinyl ether moiety can be used as a masked ketone during subsequent manipulation in the X substituent, and then hydrolyzed to ketone (equation 5.24). Alternatively, dehydrogenation of **5.54**, allows access to *para*-substituted anisoles **5.56**, and this type of process has also been used in total synthesis. Decomplexation using the more acidic reagents, e.g., $CuCl_2$ in ethanol, leads to *in situ* hydrolysis of the vinyl ether to give cyclohexenones **5.57** directly.

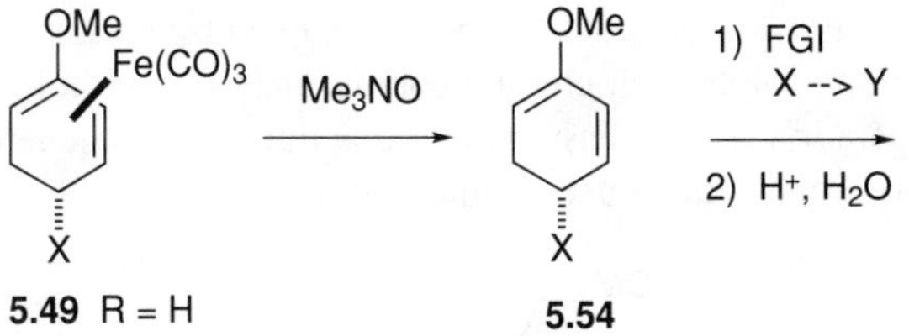

(5.24)

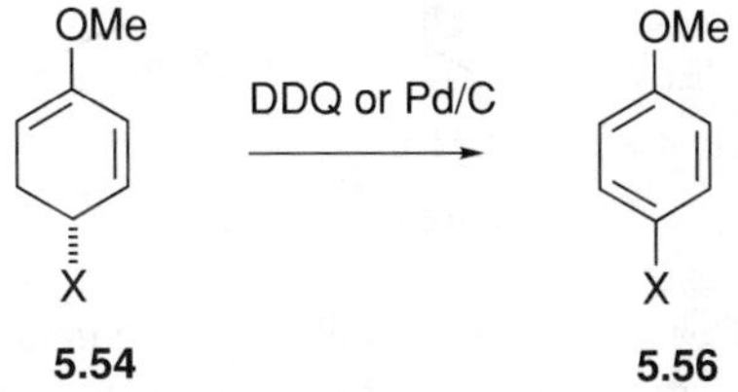

(5.25)

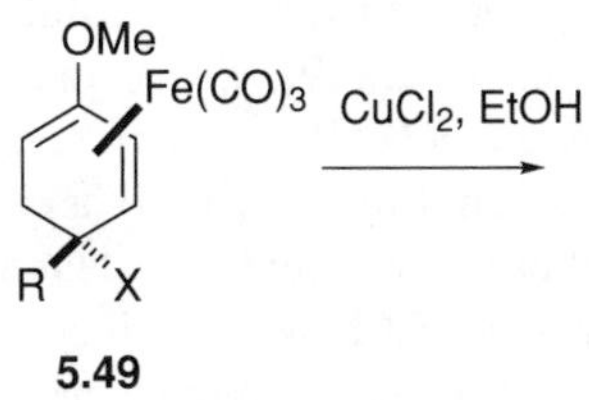

(5.26)

Complexes **5.4**, **5.7**, **5.9**, **5.11**, as well as other unsymmetrically substituted dienyls, all have planar chirality. Since nucleophile additions to the these molecules is stereospecific, and since decomplexation can be effected without loss of stereochemical integrity, there exists an opportunity for asymmetric synthesis applications, provided the starting complexes can be prepared in optically pure form. Despite the considerable potential for this methodology, only limited success has been noted with such endeavours. The ester substituted complex **5.11** has been prepared in optically pure form via resolution of the carboxylic acid **5.58** (equation 5.27)[27]. The procedure entails formation of the ammonium salt using (−)-1-phenylethylamine, followed by recrystallization from chloroform/acetone. The individual diastereomers thus obtained are converted into the optically pure acids **5.58A** and **5.58B**, having the indicated absolute stereochemistries (*ca* 34% yield of each). These are easily converted into the corresponding methyl esters and thence to the dienyl salt **5.11**.

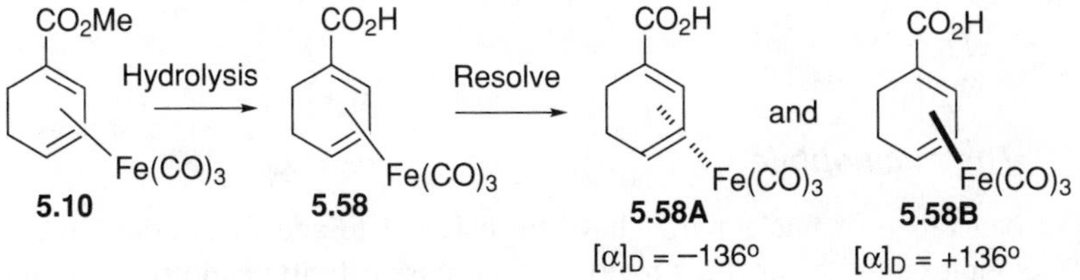

(5.27)

Resolution of the dienyl complex **5.4** has been achieved by means of addition of a chiral heteroatom nucleophile (e.g. chiral alcohols such as menthol) followed by chromatographic separation of the diastereomeric adducts and regeneration of the dienyl complexes[28]. This is illustrated in equation 5.28.

OMe, $Fe(CO)_3$, **5.4** —[R*OH; $(Pr^i)_2NEt$, CH_2Cl_2]→ OMe, $Fe(CO)_3$, OR*, **5.59** —[1) chromatography; 2) HPF_6]→ (+)-**5.4** and (−)-**5.4** (5.28)

Although these approaches provide both enantiomers of the dienyl complex, contemporary organic synthesis calls for more efficient methods. Asymmetric synthesis of these complexes has not been very successful, the best approach to the date being that reported by Birch *et al.*[29]. Based on the ability of an enone-$Fe(CO)_3$ complex to behave as an $Fe(CO)_3$ transfer reagent, it was argued that complexes in which the enone is chiral might lead to asymmetric induction during such transfer. The enones chosen for study, pulegone (**5.60**) and (−)-3β-(acetyloxy)pregna-5,16-diene-20-one (**5.61**) were each converted into their $Fe(CO)_3$ complexes *in situ*, by treatment with $Fe_2(CO)_9$. When carried out in the presence of, for example, diene **5.62** (see Chapter 4), modest degrees of asymmetric induction were observed during the formation of (+)- and (−)-**5.8** (equations 5.29 and 5.30).

5.60 + OMe, Me, **5.62** —[$Fe_2(CO)_9$, pet. ether, 40 °C, 270h; 79% yield, 10% d.e.]→ OMe, $Fe(CO)_3$, Me, (+)-**5.8** (5.29)

AcO, O, **5.61** —[**5.62**, $Fe_2(CO)_9$, PhH, 65 °C, 88h; 26% yield, 43% d.e.]→ OMe, $Fe(CO)_3$, Me, (−)-**5.8** (5.30)

5.1.2.2 The nucleophile

A very broad range of nucleophiles have been found to add to cyclohexadienyl-$Fe(CO)_3$ complexes with excellent regio- and stereoselectivity, and usually in high yield. These range from amines to alkoxides, enolates, allylsilanes and activated aromatics. There is virtually no nucleophilic type that has not given a successful result, and this area has been reviewed thoroughly[30–34]. With some types of nucleophile, initial problems have been overcome by modifying reaction conditions such as solvent, or by changing the reactivity of the dienyl complex by substituting a CO spectator ligand with triphenylphosphine. For example, in the early stages of research on these complexes, it was found that their reactions with Grignard reagents and alkyllithiums did not give satisfactory yields of alkylation product[35]. The major problem is that an electron transfer from nucleophile to dienyl cation occurs to give a putative dienyl free radical which dimerizes to give complex **5.63** (equation 5.31). This was later overcome by moderating the reactivity of the nucleophile, and it was shown that dialkylzinc, dialkylcadmium, and alkylcuprate reagents[36,37] give satisfactory yields of alkylation products **5.64**. More recently, it has been shown that the choice of solvent is critical in the reactions with alkyllithiums[38]. All of the earlier alkylation attempts had been carried out using diethyl ether or tetrahydrofuran as solvent (in which, incidentally, the dienyl complexes are insoluble). It was found that when the alkyllithium solution (usually in ether or hexane) was added to a stirred solution of the dienyl complex in *methylene chloride*, cooled to −78°C, clean alkylation occurred to give **5.64** (equation 5.32). This technique fails when vinylmagnesium bromide is reacted with complex **5.2**, and reductive dimerization is still the major reaction. The use of triphenylphosphine(dicarbonyl)cyclohexadienyliron complex **5.65** completely remedies this problem and the alkylation product **5.66** is obtained in excellent yield[39].

$Fe(CO)_3$ BF_4^- **5.2** —(RLi or RMgX, Et_2O or THF)→ $(CO)_3Fe$ … $Fe(CO)_3$ **5.63** (major product) + $Fe(CO)_3$ R **5.64** (minor) (5.31)

5.2 —(R_2Zn *or* R_2Cd *or* R_2CuLi *or* RLi, CH_2Cl_2, −78 °C; *ca* 30 – 90% yields)→ **5.64** (5.32)

$Fe(CO)_2PPh_3$ PF_6^- **5.65** —($CH_2{=}CHMgBr$, CH_2Cl_2, −78 °C, 98%)→ $Fe(CO)_2PPh_3$ $CH{=}CH_2$ **5.66** (5.33)

Quite unreactive nucleophiles give addition products, reflecting the highly reactive nature of the dienyl complexes. For example, the enol form of ketones will add to give adducts such as **5.67**, and the reaction is readily carried out by simply refluxing the complex in a mixture of, for example, acetone and ethanol[35] (equation 5.34). Activated aromatics undergo alkylation. For anilines, two pathways are possible, *N*-alkylation or *C*-alkylation, and either can be obtained, depending on reaction conditions. At room temperature, using acetonitrile as solvent, the product of *N*-alkylation (**5.69**) is obtained. When the reaction is carried out in hot acetonitrile, *C*-alkylation (**5.70**) is observed. This is an excellent example of thermodynamic vs. kinetic control, as evidenced by the fact that **5.69** is converted into **5.70** by refluxing in acetonitrile[40,41].

R Fe(CO)3 + BF4- ; CH3COCH3, EtOH, reflux ; R = H or MeO ; R Fe(CO)3 CH2COCH3 **5.67** (5.34)

R Fe(CO)3 + BF4- ; R = H or MeO ; + NH2 Me **5.68** ; CH3CN, rt ; R Fe(CO)3 HN Me **5.69** (5.35)

5.2 ; **5.68**, CH3CN, Δ ; Fe(CO)3 NH2 Me **5.70** (5.36)

One exception to the generality regarding compatible nucleophiles is the disubstituted complex **5.9**, as well as complexes in which the methyl group is replaced by longer alkyl chains. While there is a fairly broad range of nucleophiles that add to the dienyl to give complexes of general structure **5.49**, this behavior is not observed for highly basic nucleophiles. Instead, deprotonation of the methyl (or alkyl) group occurs to give the η^4-triene complex **5.71** (equation 5.37).

Nucleophiles that add to the dienyl include stabilized enolates, such as malonic esters and β-keto ester enolates, and relatively soft tin enolates. Grignard reagents, organolithiums and non-stabilized lithium enolates all promote deprotonation. Previously, all attempts to effect C–C bond formation using organocuprates also led to deprotonation, but recently it has been found that low order vinylcopper derivatives (vinylcopper-Me_2S) will deliver a vinyl group to the methyl-substituted terminus of **5.9**(b) (equation 5.38)[42]. This has led to a short synthesis of the tamoxifen analog **5.73**.

OR, $Fe(CO)_3$, +, Me — Basic nucleophiles → OR, $Fe(CO)_3$, CH_2 (5.37)

5.9a R = Me
5.9b R = Pr^i

5.71

OR, Br — 1) Bu^nLi, THF, −78 °C; 2) $Cu(Me_2S)Br$; 3) **5.9b** → OPr^i, $Fe(CO)_3$, Me, RO **5.72**

$CuCl_2$, EtOH ↓

O, Me, RO **5.73** (5.38)

The following equations give a listing of nucleophile types that add to cyclohexadienyliron complexes. This is not an exhaustive survey, but is meant to give the reader a feel for those reactions that work well. Examples of nucleophiles and complexes that have been developed specially for total synthesis applications are also discussed in the next section.

$R = H$ or OMe; BF_4^- or PF_6^-; morpholine → **5.74** (5.39)

$R = H$ or OMe; BF_4^- or PF_6^-; Alkoxides → OR' **5.75** (5.40)

$R = H$ or OMe; BF_4^- or PF_6^-; R'–C_6H_4–NH_2, 25 °C → NHAr **5.76** (5.41)

$R = H$ or OMe; BF_4^- or PF_6^-; R', R''–C_6H_3–NH_2, Heat → H_2N, R', R'' **5.77** (5.42)

$Fe(CO)_3$ + furan, Δ → **5.78** (5.43)

$Fe(CO)_3$ + N-R indole →(Δ) **5.79** (5.44)

$Fe(CO)_3$ + MeO, OMe →(Δ) H, OMe, OMe **5.80** (5.45)

R, $Fe(CO)_3$ + $NaCH(CO_2Me)_2$ →(THF, 0°C; 95–100%) R, $Fe(CO)_3$, $CH(CO_2Me)_2$ **5.81** (5.46)

R = H or MeO

MeO, $Fe(CO)_3$, Me, PF_6^- + NaO, MeO_2C →(THF, 0°C; 98%) $Fe(CO)_3$, O, MeO, Me, CO_2Me (5.47)

5.82 (1:1 diastereomers)

Me, PF_6^-, MeO, $Fe(CO)_3$ + O, O, MeO, $OSiMe_3$ → Me CO_2Me, O, O, MeO, $Fe(CO)_3$ (5.48)

5.83

$R'CH_2COMe$, Δ (5.49)

R = H or MeO

$CH(R')COMe$

5.84

$Fe(CO)_3$ BF_4^-

EtOH, Δ (5.50)

5.85

1) 2) H_2O (5.51)

5.86

1) OSiMe$_3$ –OSiMe$_3$ 2) H^+, MeOH (5.52)

5.87

$NaBH_4$ (5.53)

5.88

R'_2Cd, R'_2Zn, or R'_2CuLi (5.54)

5.89

Fe(CO)3 BF4- + (Me-CH=CH-CH2)2Cd → Fe(CO)3, Me, **5.90** (5.55)

R, Fe(CO)3 BF4- + SiMe3, Heat → R, Fe(CO)3, **5.91** (5.56)

R, Fe(CO)3 BF4- + -CH2NO2 → R, Fe(CO)3, CH2NO2, **5.92** (5.57)

MeO, Fe(CO)3 BF4- + [R'3B—≡—R]- → MeO, Fe(CO)3, R', BR'2, R, **5.93** (5.58)

Fe(CO)3 BF4- + PR3 → Fe(CO)3, +PR3 BF4-, **5.94** (5.59)

Fe(CO)3 BF4- + P(OMe)3, MeOH → Fe(CO)3, PO(OMe)2, **5.95** (5.60)

5.1.3 Synthetic applications of cyclohexadienyliron complexes

5.1.3.1 Strategic considerations

These complexes may be used as the electrophilic partners in polar reactions, allowing the formation of C–C and C–X (heteroatom) bonds of key significance in the synthesis of a variety of natural products. From the earlier discussion it should be apparent that the ultimate construction of substituted cyclohexadienes, cyclohexenones and aromatic compounds can be contemplated, depending upon subsequent decomplexation and/or dehydrogenation tactics. As a reminder, three sets of transformations are shown in equations 5.61–5.63.

MeO, $Fe(CO)_3$, PF_6^-, R — Nucleophile → MeO, $Fe(CO)_3$, R', R — [O] / H^+, H_2O → O, R', R (5.61)

$Fe(CO)_3$, PF_6^-, R — Nucleophile → $Fe(CO)_3$, R, R' — Decomplex → R, R' (5.62)

R, $Fe(CO)_3$, PF_6^- — Nucleophile → R, $Fe(CO)_3$, R' — Decomplex / Dehydro-genate → R, R' (5.63)

Such transformations can be of key significance, since cyclohexenones and cyclohexadienes provide useful functionality on which to build the remaining parts of a target molecule, or they form an integral part of such targets, while substituted aromatic molecules that are not readily prepared using standard methodology are potentially accessible via the transformations shown in equation 5.63. For the purposes of retrosynthetic analysis, therefore, we can imagine a collection of useful synthons and their corresponding synthetic equivalents, as shown in Fig. 5.2.

This concept opens up a new set of possible bond disconnections that can be visualized in a retrosynthetic analysis of a target molecule. Synthons **5.96**, **5.99** (R = electron donor), **5.100** (R = EWG), **5.101** (R = MeO) and **5.102** (R = EWG) are interesting and useful because they represent examples of *umpolung*[43] that are not readily accessible *via* traditional organic synthesis methodology. Synthon **5.97** is also interesting, as it provides the equivalent of a conjugate addition to cyclohexa-2,5-dien-3-one, which can otherwise only be accessed by a stepwise

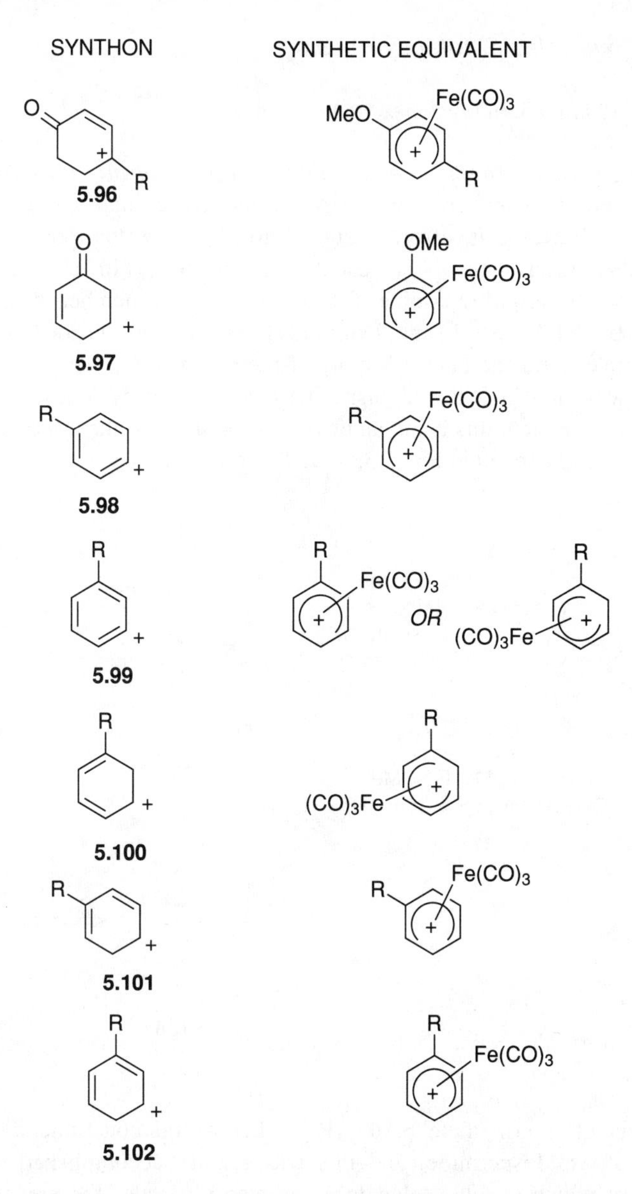

FIG. 5.2. Synthons and synthetic equivalents available as cyclohexadienyliron complexes.

procedure (conjugate addition to cyclohexenone, followed by regiocontrolled introduction of the double bond). Furthermore, there is virtually no limitation on the type of substituent that can be set in place using complex **5.5**. Thus, using the guidelines for synthons shown in Fig. 5.2 we can make strategic decisions relating to complex molecular synthesis.

5.1.3.2 *Examples of total synthesis*

(a) Uses of synthon **5.96** *(cyclohexenone δ-cation)*

(i) Aspidosperma alkaloids - an exercise in regiocontrol Two Aspidosperma alkaloids were targeted for this work, which constituted the first successful multistep total synthesis effort designed to illustrate the potential utility of cyclohexadienyliron cations as reactive intermediates in complex molecular synthesis[44–46]. Aspidospermine (**5.103**), the simplest member of the group, had been synthesized by Stork and Dolfini[47], and it was decided to employ an identical strategy for the later assembly of the ring system, the point of departure being the way in which the requisite 4,4-disubstituted cyclohexenone precursor **5.108** was constructed, this being an obvious candidate for the use of synthon **5.96**. The strategy is illustrated in Fig. 5.3, in outline only.

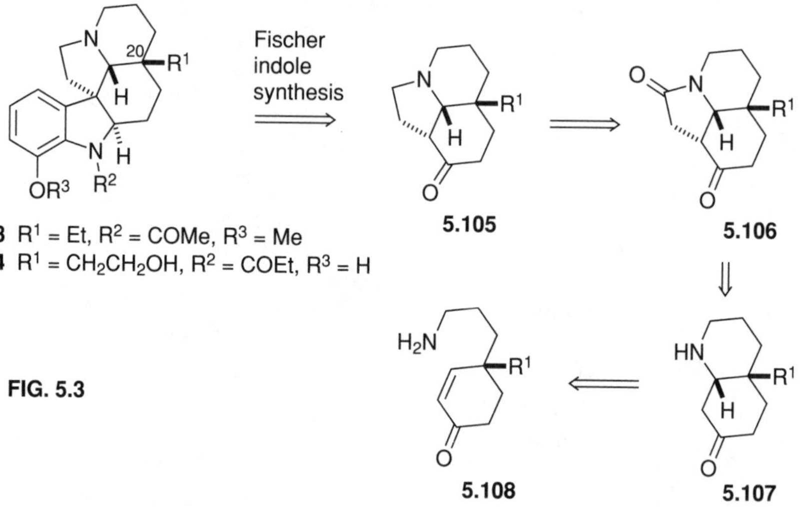

Fischer indole synthesis

5.105 **5.106** **5.107** **5.108**

5.103 R^1 = Et, R^2 = COMe, R^3 = Me
5.104 R^1 = CH_2CH_2OH, R^2 = COEt, R^3 = H

FIG. 5.3

Synthesis of intermediate **5.107** (R^1 = Et) would constitute a formal total synthesis of aspidospermine, and this was readily accomplished as shown in Scheme 5.4, which is self explanatory and also illustrates the preparation of the requisite diene complex **5.109**. The reaction of dienyl complex **5.110** with malonate enolate, followed by decarboxylation, was to set in place the functionalized two-carbon residue that could be homologated to give the piperidine ring residue.

At the outset of this work there were several unknowns. The regioselectivity during nucleophile additions to complexes of type **5.110**, which carried a terminal alkyl substituent larger than methyl, had not previously been examined.

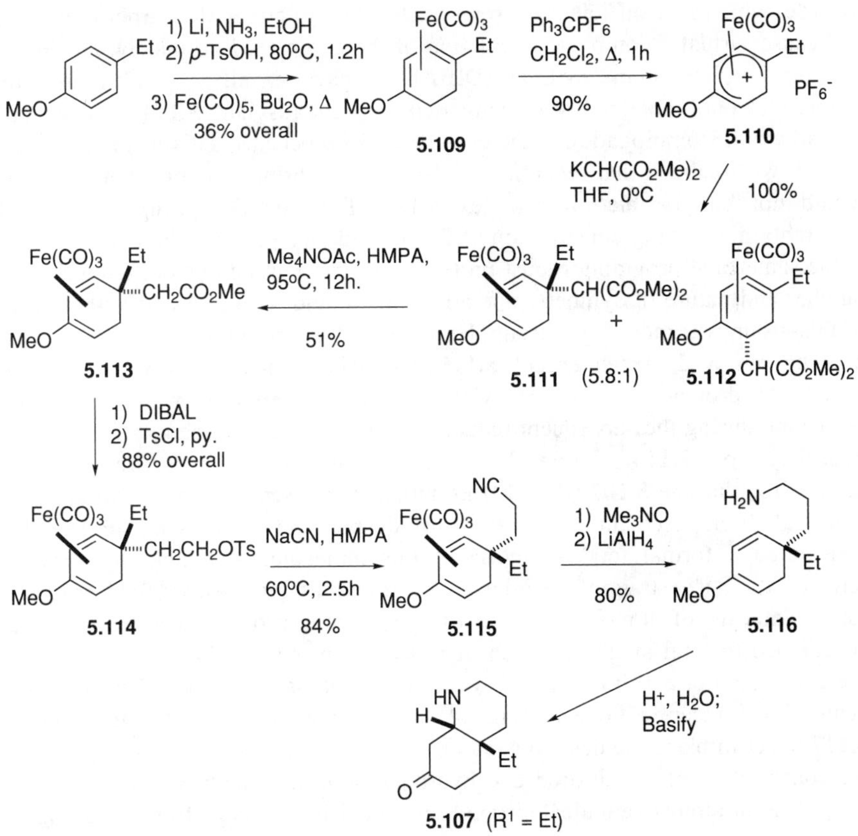

SCHEME 5.4. Formal total synthesis of (±)-aspidospermine.

The fact that similar complexes derived from hexahydronaphthalenes reacted with malonate with excellent regioselectivity was encouraging[48], but in fact the reaction of **5.110** with $NaCH(CO_2Me)_2$ was quite discouraging, giving a 4.6:1 mixture **5.111** and **5.112**. During the course of this work it was found that the nature of the enolate countercation significantly affected the regioselectivity of the reaction[49]. For example, $LiCH(CO_2Me)_2$ reacted with **5.110** to give a 3.0:1 mixture of **5.111** and **5.112**, whereas $KCH(CO_2Me)_2$ gave the 5.8:1 mixture shown in Scheme 5.4, from which pure crystalline **5.111** was obtained in 78% yield from **5.110** by crystallization from pentane. At this stage the result was deemed good enough, and the remaining unknowns were investigated. Decarboxylation of malonate esters using supernucleophile conditions[50,51] gave **5.113**, albeit in low yield (51%), but this procedure was improved in later work. Reduction of the ester **5.113** could not be effected with lithium aluminum hydride, which is a sufficiently strong nucleophile to attack the carbonyl ligands and cause virtually complete destruction of the complex. Fortunately, an excess of di-*iso*butylaluminum hydride (DIBAL-H) gave excellent results, and the primary alcohol obtained was converted into its tosylate **5.114** in excellent overall yield. Homologation to the cyanide **5.115** proceeded satisfactorily. At this point it was realized that reduction of the nitrile to primary amine using $LiAlH_4$ would not be possible on complex **5.115**. The $Fe(CO)_3$ group had served admirably in allowing construction of the desired quaternary carbon center while at the same time providing robust protection of the masked cyclohexenone. The number and nature of synthetic operations that could be performed without loss of this group was indeed gratifying. However, it had served its purpose and must now be removed. Treatment of **5.115** with trimethylamine-*N*-oxide gave the decomplexed dienol ether in 88% yield, in which the enol ether group provided protection during the subsequent reduction of the nitrile, leading to **5.116**. Acidic hydrolysis of **5.116,** followed by appropriate work up, gave the *cis*-decahydroquinoline **5.107** (R^1 = Et) as anticipated based on Stork's work. Since this had been used in the earlier aspidospermine synthesis, Scheme 5.4 constituted a formal total synthesis of this molecule. Although this was, in retrospect, a fairly straightforward exercise, it laid the groundwork for a proposed total synthesis of limaspermine (**5.104**) a natural product that had not yet succumbed to total synthesis. Also, it posed the problem of regiocontrol, since the conversion of **5.110** into **5.111** was not as good as was hoped. The first total synthesis of racemic limaspermine[44] used a malonate addition to complex **5.117** to set in place the desired quaternary carbon with functionality appropriate for construction of the desired decahydroquinoline intermediate. However, this was a task of almost Herculean proportions, requiring 28 steps from the aromatic

Fe(CO)3, (CH2)3NPhthal, PF6⁻, MeO — **5.117** —[KCH(CO2Me)2, 68%]→ PhthalN, Fe(CO)3, CH(CO2Me)2, MeO — **5.118** (5.64)

(Phthal = Phthalimido)

starting material (*p*-methoxycinnamic acid) to the final product. Also, the reaction of **5.117** with dimethyl malonate enolates suffered the same regioselectivity problems as for **5.110**, $KCH(CO_2Me)_2$ giving only 68% yield of the desired complex **5.118**. Therefore, we shall step back at this point and discuss how improvements in regioselectivity for this reaction were developed. A more efficient total synthesis of racemic limaspermine that emerged as a result of these studies will be described based on the route shown in Scheme 5.4.

It was realized that one possible method for exerting greater regiocontrol

$Fe(CO)_3$ Et Pr^{i}O **5.119** — Ph_3CPF_6, CH_2Cl_2, Δ, 1h, 100% → $Fe(CO)_3$ Et + PF_6^- Pr^{i}O **5.120** — $KCH(CO_2Me)_2$, THF, 0°C, 97% → $Fe(CO)_3$ Et $CH(CO_2Me)_2$ Pr^{i}O **5.121** (5.65)

during nucleophile additions to **5.110** would be to replace the methoxy directing group with a more sterically demanding ether. Since t-butyl ethers were potentially problematic during acid-catalyzed isomerization of the cyclohexa-1,4-dienes, required for the initial preparation of the diene-$Fe(CO)_3$ complexes, it was decided to investigate the use of *iso*propyl ethers. The main concern here was whether hydride abstraction could still be carried out regioselectively, since attack by the bulky trityl cation might be more difficult at the methylene next to an *iso*propoxy group. In the event, this reaction worked well, complex **5.119** being converted into **5.120** in essentially quantitative yield[52]. Reaction of **5.120** with $KCH(CO_2Me)_2$ gave the desired diene complex **5.121** as a *single regioisomer* in 97% yield. Thus, the problem of regiocontrol encountered in Scheme 5.4 was solved, and the stage was set for an assault on the more challenging target limaspermine.

We shall describe the total synthesis of limaspermine that is a combination of two pieces of work, the later stages being adapted from the synthesis using complex **5.118**, whereas the earlier stages are based on a more efficient construction of the decahydroquinoline system from complex **5.123**, and this is summarized in Scheme 5.5. Mostly this scheme is straightforward. Surprisingly, inclusion of a methyl ether in the side chain of **5.123** (vs. **5.120**) does have an effect on the subsequent nucleophile addition, which now gives a *ca* 9:1 mixture in favor of **5.124.** As in the aspidospermine synthesis, the $Fe(CO)_3$ group serves as a dienol ether protecting group during a number of subsequent transformations, all of which are surprisingly well behaved. The conversion of **5.126** into limaspermine is virtually identical to the sequence employed in Stork's aspidospermine synthesis, with the exception of methyl ether deprotection (see **5.130** to **5.104**) which was a rather difficult and low-yielding operation. The synthesis is a fairly respectable 22 steps from the aromatic precursor.

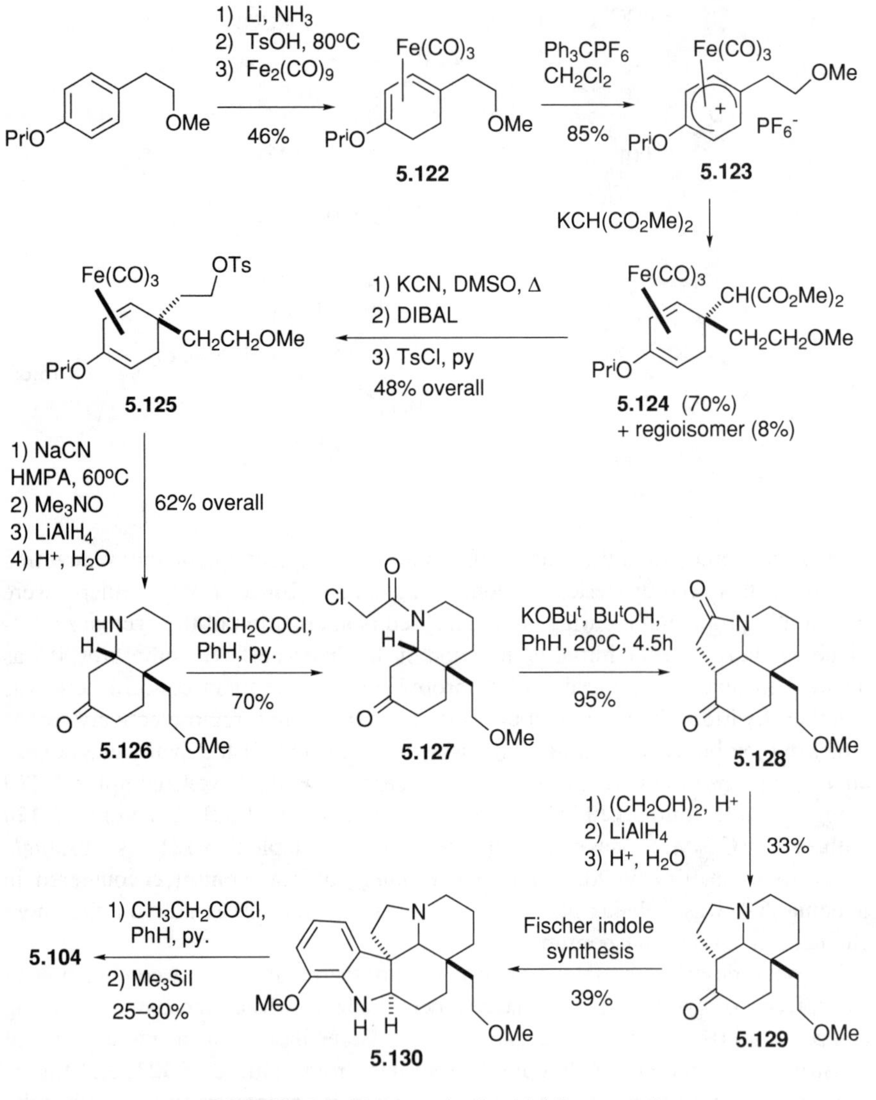

SCHEME 5.5. Total synthesis of (±)-limaspermine.

(ii) Total synthesis of (±)-trichodiene and (±)-trichodermol; an exercise in nucleophile selection The alkaloid syntheses described above were successful in demonstrating that cyclohexadienyliron complexes can be used in multistep synthesis. However, they borrow heavily from prior art and offer no real advantages over traditional "organic" approaches. The potential utility of this methodology would be better showcased by total synthesis of a complex natural product in which bond formations and functional group interconversions are

accomplished that are otherwise impossible or very difficult. Furthermore, the resulting synthesis should be better, in terms of operational efficiency, than any previous "organic" approach. With this in mind, a study in the area of trichothecene synthesis was undertaken[53]. The ultimate target molecules chosen were trichodermol (**5.131**), one of the simplest members of the class, and trichodiene (**5.132**), the biogenetic precursor of all members[54–56]. We shall first outline the organoiron routes to these molecules and then make a comparison with other total syntheses.

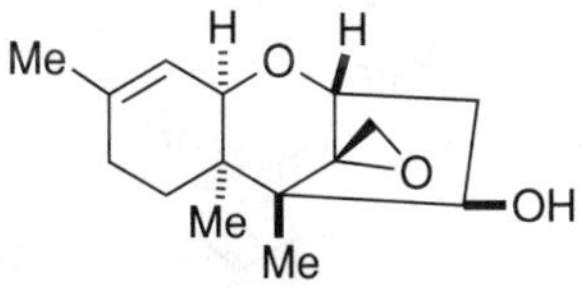

Me H O H O OH Me Me

5.131

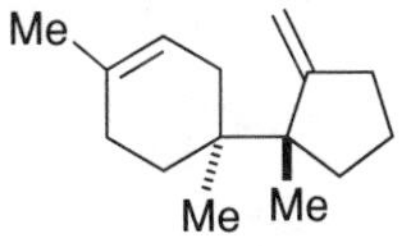

Me Me Me

5.132

It was envisioned that complex **5.9** would provide the six-membered carbocyclic ring of **5.131** and **5.132**. A rapid entry into the carbon skeleton could best be effected if a bond connection could be made between **5.9** and a 2-methyl cyclopentanone enolate. Preliminary studies indicated that such regiocontrolled coupling could indeed be achieved using the enolate from methyl 2-oxo-cyclopentane carboxylate, which gave complex **5.82** (equation 5.47) in near quantitative yield. The main drawback was the lack of diastereoselectivity, although functionalization of the five-membered ring could be accomplished, and this resulted in the construction of some unnatural analogs of the target molecules. However, in order to access the trichodiene and trichodermol structures, it would be necessary to reduce the sterically hindered ester to a methyl group, a difficult multistep operation. Consequently, it was desirable to directly couple complex **5.9** with enolates of general structure **5.133**. The problem here is that lithium enolates are quite basic, and in this particular reaction a virtually quantitative conversion of **5.9** into **5.71** occurs (see equation 5.37). Clearly, a less reactive enolate was required. Enolsilanes had been shown[57] to react with simpler cyclo-hexadienyliron complexes (equation 5.52). Although the enolsilane **5.133** (M = $SiMe_3$) reacted satisfactorily with complex **5.4**, it was found that several days refluxing in methylene chloride, using a 20-fold excess of enolsilane was required for conversion of **5.9** into **5.134** in 7% yield. A practical solution to this problem was revealed using the slightly more reactive tin enolate (**5.133**, M = $SnBu_3$) which was found to react with **5.9** at *ca* −50°C to give **5.134** in 87% yield. Moreover, as an added bonus a 5:1 mixture of diastereomers was obtained, in favor of **5.134**, which has relative stereochemistry corresponding to the trichothecenes. In fact, these cyclopentanone tin enolates are the only prochiral nucleophiles to show any stereoselectivity during their reactions with complex **5.9**. Nature had indeed been very kind. Almost the entire trichodiene molecule had been set in place in a single reaction. Scheme 5.6 illustrates the conversion of complex **5.134A** to racemic

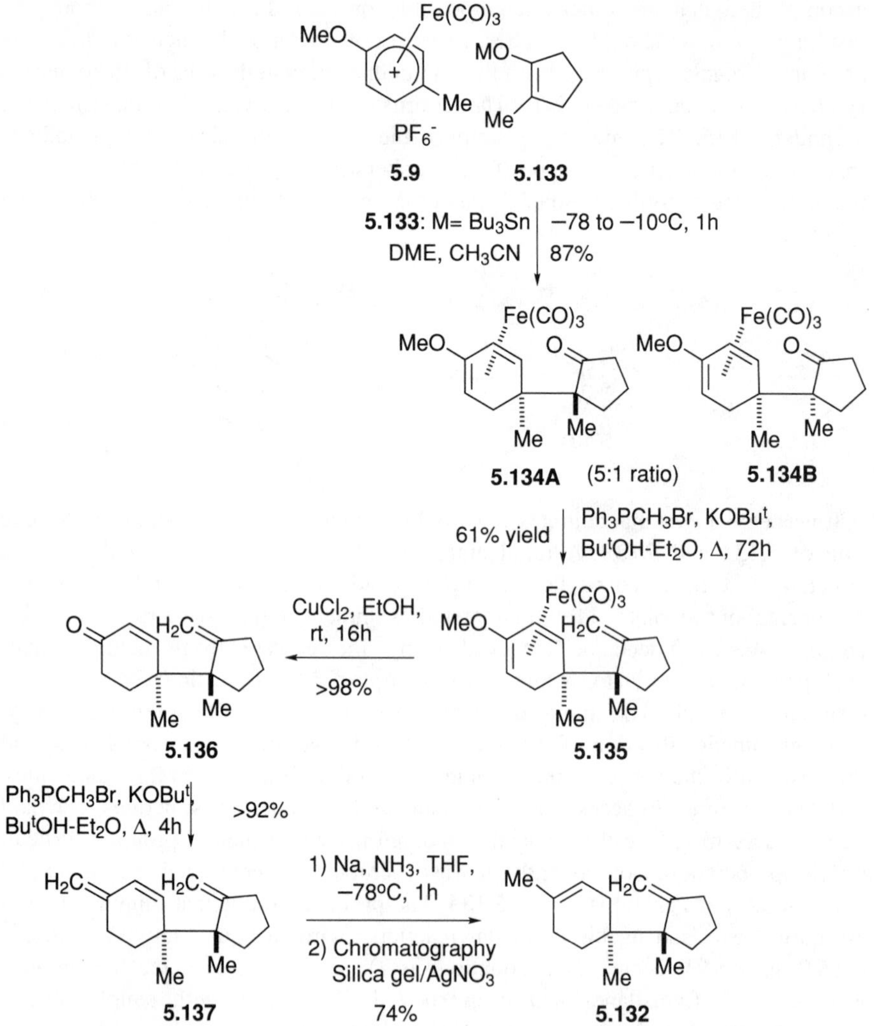

$Fe(CO)_3$ MeO MO Me Me PF_6^-

5.9 **5.133**

5.133: M= Bu_3Sn, DME, CH_3CN | −78 to −10°C, 1h, 87%

5.134A (5:1 ratio) **5.134B**

61% yield | Ph_3PCH_3Br, $KOBu^t$, Bu^tOH-Et_2O, Δ, 72h

5.135 → $CuCl_2$, EtOH, rt, 16h, >98% → **5.136**

Ph_3PCH_3Br, $KOBu^t$, Bu^tOH-Et_2O, Δ, 4h | >92%

5.137 → 1) Na, NH_3, THF, −78°C, 1h; 2) Chromatography Silica gel/$AgNO_3$; 74% → **5.132**

SCHEME 5.6. Total synthesis of (±)-trichodiene.

trichodermol. One of the interesting features of this synthesis is the Wittig methylenation of complex **5.134A**. This has to be done prior to decomplexation, since intramolecular aldol reactions were found to occur during any attempts to demetallate **5.134** to give the cyclohexenone derivatives. This route provides an exceptionally short (five synthetic operations from **5.9**, nine from *p*-methyl anisole) high overall-yield synthesis of trichodiene (30% from **5.9**). Bearing in mind that **5.9** is easily prepared on a scale of 250 g or more, it is probably the most convenient synthesis of trichodiene so far recorded.

With regard to the trichodermol problem, a modification of the tin enolate nucleophile was required that would allow the C-4 hydroxyl to be set in place with stereocontrol. The best OH surrogate appeared to be a phenyldimethylsilyl group, which had been shown to be convertible into OH with retention of stereochemistry at the attached carbon[58,59]. The required tin enolate was readily prepared by the route shown in Scheme 5.7, involving silylcuprate addition to

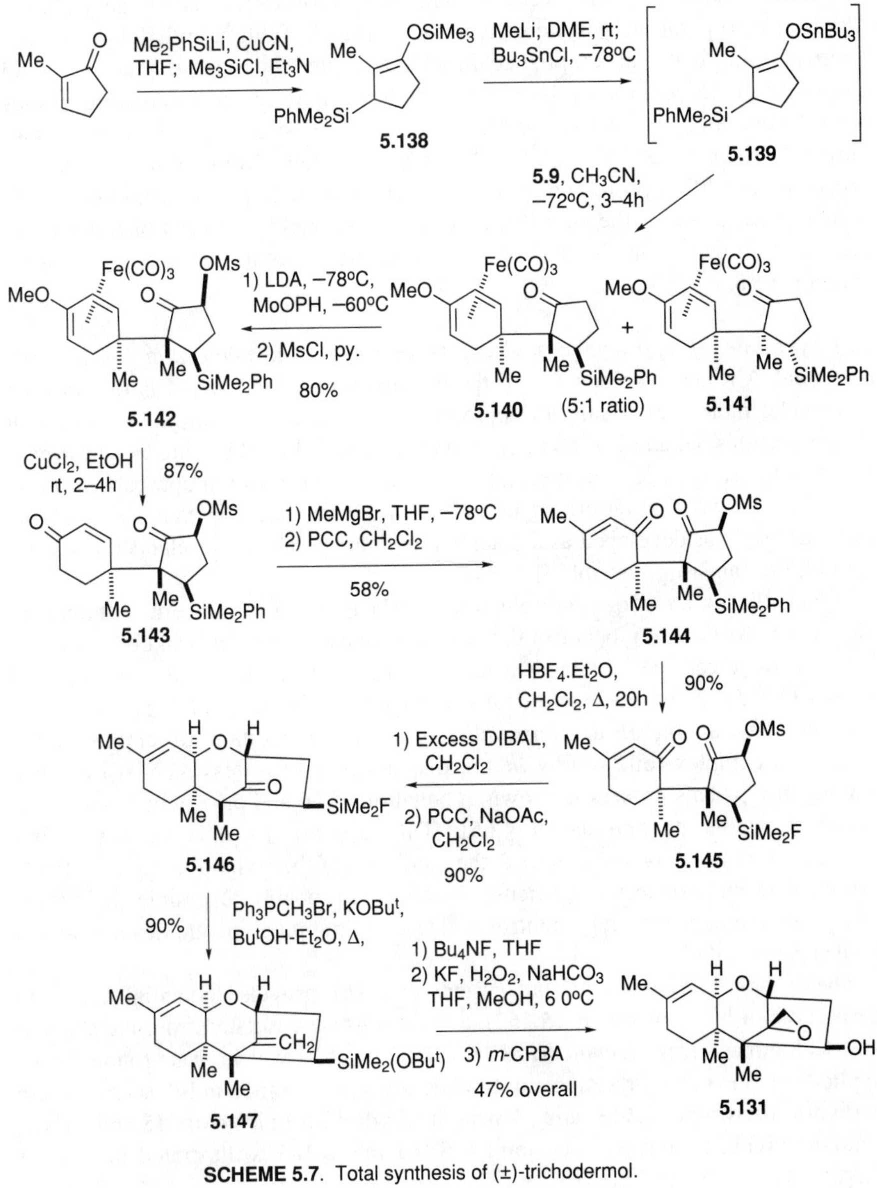

SCHEME 5.7. Total synthesis of (±)-trichodermol.

2-methylcyclopentenone, and trapping of the intermediate enolate with chlorotrimethylsilane. The enolsilane thus produced (**5.138**) could be purified by distillation and provided a convenient precursor to the reactive tin enolate **5.139**. Reaction of **5.139** with complex **5.9** was again regiospecific and diastereoselective, giving only the two isomers **5.140** and **5.141** as a 5:1 mixture, easily separated by crystallization (**5.141** is virtually insoluble in diethyl ether). All subsequent transformations were completely stereoselective and high-yielding. The trickiest operation was the conversion of the phenyldimethylsilyl group into hydroxyl, this being accomplished in an interrupted stepwise manner to avoid having to protect/deprotect the alcohol. Indeed, there are no protecting groups used throughout the entire sequence. The total synthesis of trichodermol outlined in Scheme 5.7 requires 13 synthetic steps from complex **5.9** and proceeds in 9.8% overall yield (17 steps from *p*-methylanisole, approx. 5.8% overall yield). This is the shortest, highest overall-yield synthesis of this natural product recorded to date[60,61] and it clearly illustrates the advantages of using synthon **5.96**.

(iii) *Synthesis of spirocyclic molecules: inter- vs. intramolecular nucleophile addition* An obvious extension of the chemistry discussed above is to construct spirocyclic molecules. Using an appropriately functionalized terminal substituent (R in structures **5.4** and **5.9**) addition of nucleophile followed by ring closure would result in the spirocycle. An intermolecular nucleophile addition approach to these ring systems was first described in 1979[62], and is shown in Scheme 5.8. This methodology was developed as a potential approach to natural products that have a spiro[4.5]decane ring system[63].

Again, the rather poor regioselectivity (4:1) is of some concern, although in view of the work on regiocontrol that appeared subsequently, it is likely that much better results can be obtained using *iso*propyl ether dienyl derivatives and $KCH(CO_2Me)_2$ as the nucleophile. An extension of this methodology allows the one-pot construction of azaspirocyclic systems, by addition of amine to the dienyliron complex followed by *in situ* displacement of a (less reactive) tosylate leaving group. This process is shown in equations 5.66 and 5.67, which show that excellent yields are produced[64,65]. This appears to indicate that better regioselectivity is obtained during the addition of benzylamine to the dienyl cation. However, since this particular reaction is probably reversible, it is more likely that the enhanced regiocontrol is due to a "trapping" of the desired adduct by its spiroannulation.

The azaspiro[5.5]undecane ring system of **5.161** provides potential access to derivatives of histrionicotoxin (**5.162**), a toxic substance isolated from the skins of the Colombian frog *Dendrobates histrionicus*[66], which has found some application in neurophysiological studies. Simpler compounds, such as the perhydro derivative **5.163** and depentylperhydrohistrionicotoxin (**5.164**) show similar activity. Conversion of complex **5.161** into **5.164** is illustrated in Scheme 5.9[67].

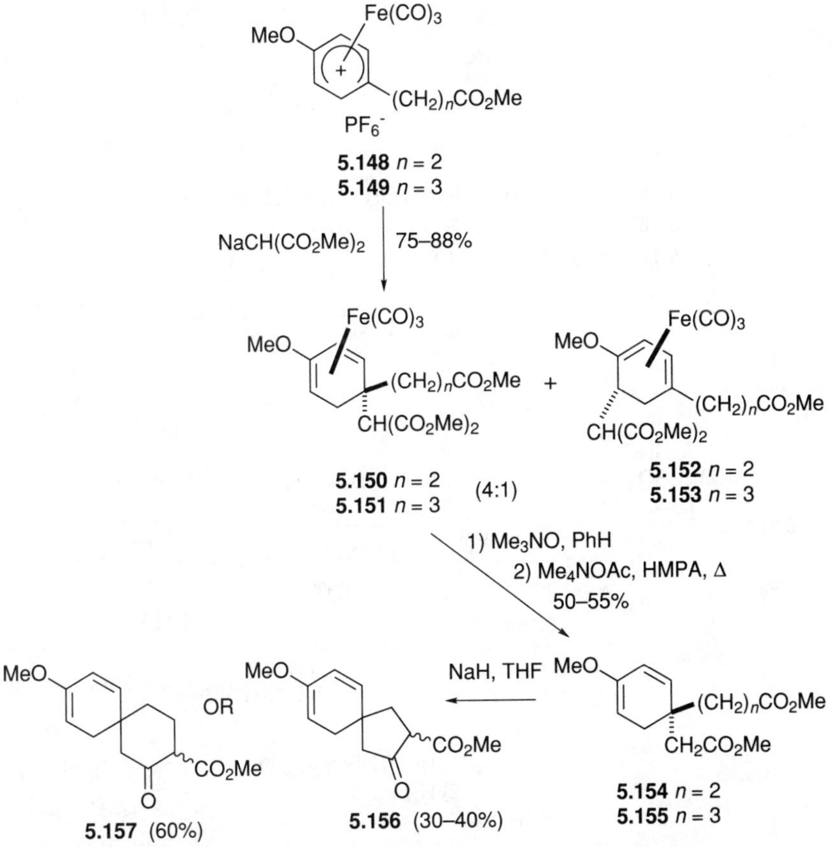

$Fe(CO)_3$
MeO
$(CH_2)_nCO_2Me$
PF_6^-

5.148 $n = 2$
5.149 $n = 3$

$NaCH(CO_2Me)_2$ 75–88%

$Fe(CO)_3$ MeO $(CH_2)_nCO_2Me$ $CH(CO_2Me)_2$ + $Fe(CO)_3$ MeO $CH(CO_2Me)_2$ $(CH_2)_nCO_2Me$

5.150 $n = 2$ (4:1) **5.152** $n = 2$
5.151 $n = 3$ **5.153** $n = 3$

1) Me_3NO, PhH
2) Me_4NOAc, HMPA, Δ
50–55%

MeO $(CH_2)_nCO_2Me$ CH_2CO_2Me

5.154 $n = 2$
5.155 $n = 3$

NaH, THF

MeO CO_2Me O

5.156 (30–40%)

OR

MeO O CO_2Me

5.157 (60%)

SCHEME 5.8. Intermolecular nucleophile addition approach to spirocyclic molecules.

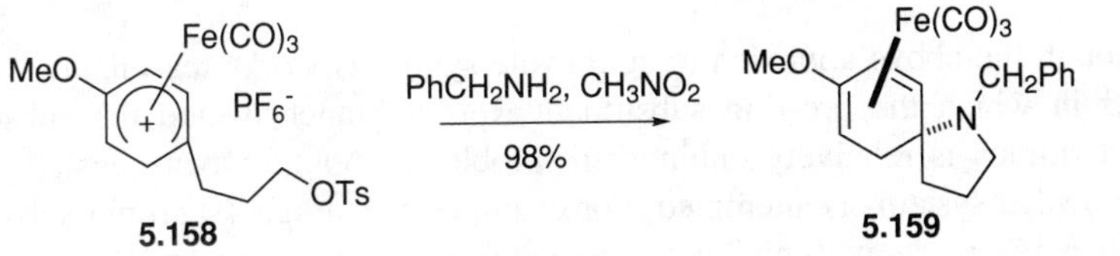

$Fe(CO)_3$ MeO PF_6^- OTs

5.158

$PhCH_2NH_2$, CH_3NO_2
98%

$Fe(CO)_3$ MeO CH_2Ph N

5.159 (5.66)

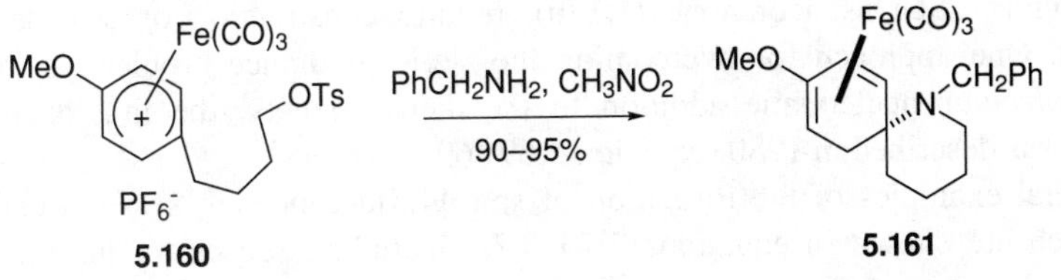

$Fe(CO)_3$ MeO OTs PF_6^-

5.160

$PhCH_2NH_2$, CH_3NO_2
90–95%

$Fe(CO)_3$ MeO CH_2Ph N

5.161 (5.67)

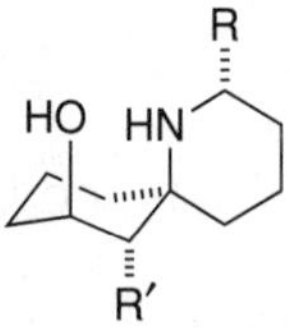

R
HO HN
R′

5.162 R = $CH_2CH\overset{c}{=}CHC\equiv CH$, R′ = $CH\overset{c}{=}CHC\equiv CH$
5.163 R = $(CH_2)_4CH_3$, R′ = $(CH_2)_3CH_3$
5.164 R = H, R′ = $(CH_2)_3CH_3$

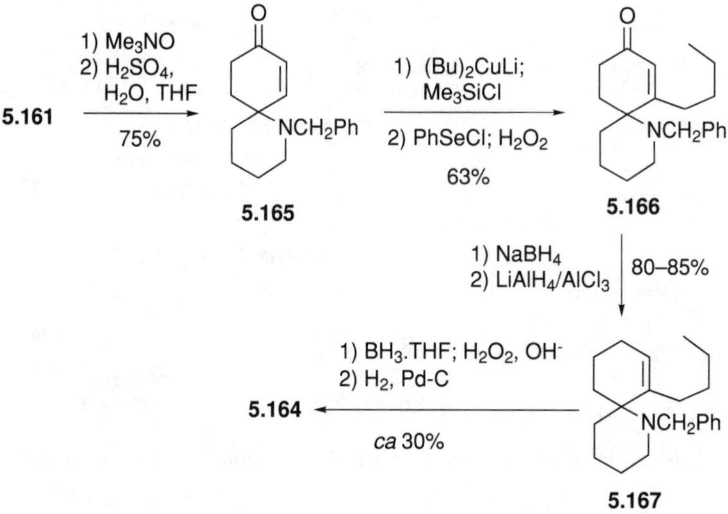

5.161 → [1) Me_3NO; 2) H_2SO_4, H_2O, THF; 75%] → **5.165** (NCH₂Ph) → [1) $(Bu)_2CuLi$; Me_3SiCl; 2) PhSeCl; H_2O_2; 63%] → **5.166** (NCH₂Ph) → [1) $NaBH_4$; 2) $LiAlH_4/AlCl_3$; 80–85%] → **5.167** (NCH₂Ph) → [1) BH_3.THF; H_2O_2, OH^-; 2) H_2, Pd-C; *ca* 30%] → **5.164**

SCHEME 5.9. Synthesis of (±)-depentylperhydrohistrionicotoxin.

Although the above approach to spirocycle synthesis works reasonably well for systems in which the pro-ring substituent R is unbranched, and the substituted dienyl terminus is relatively unhindered, problems abound when construction of more crowded systems is attempted. For example, the simple *iso*propyl substituted complex **5.168** reacts with malonate enolate *exclusively* at the unsubstituted dienyl terminus to give **5.169**[68]. A number of naturally occurring spirocyclic compounds, such as acorenone (**5.170**), require construction of crowded spiro centers. One approach for overcoming the steric hindrance problem is to effect *intramolecular* nucleophile addition to the dienyl moiety, the first example of which was described in 1980 (equation 5.69)[69].

Several examples of the formation of spiro[4.5]decane ring systems using this approach are shown in equations 5.70–5.72, including successful intramolecular

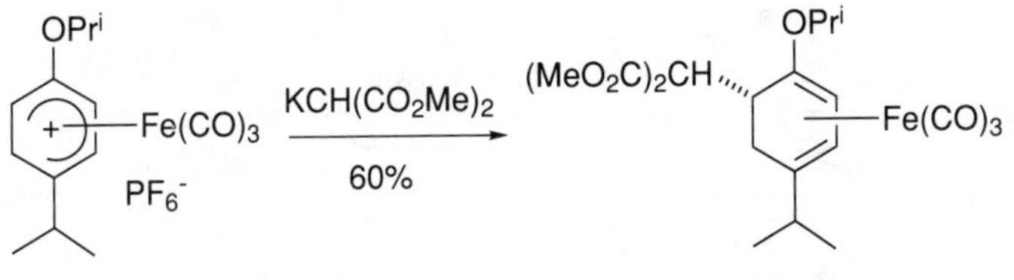

OPr[i] ; $Fe(CO)_3$; PF_6^- — $KCH(CO_2Me)_2$, 60% → $(MeO_2C)_2CH$; OPr[i] ; $Fe(CO)_3$ (5.68)

5.168 **5.169**

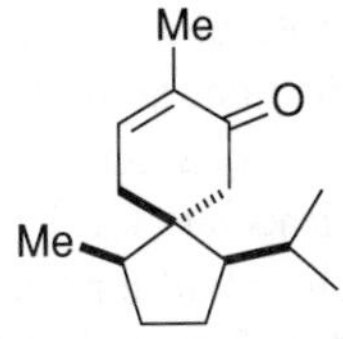

Me ; O ; Me

5.170

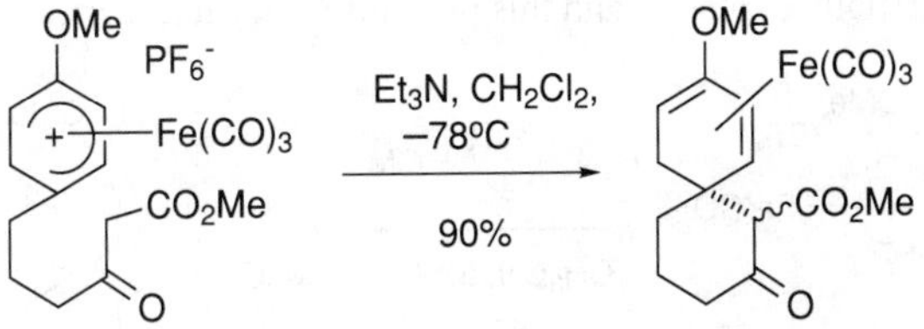

OMe ; PF_6^- ; $Fe(CO)_3$; CO_2Me ; O — Et_3N, CH_2Cl_2, −78°C, 90% → OMe ; $Fe(CO)_3$; CO_2Me ; O (5.69)

5.171 **5.172**

additions to sterically hindered dienyls[68]. It should be noted that this method is limited to the use of readily enolizable side chains, the threshold being the malonate derivative **5.173**, base treatment of which gives only the exocyclic alkene derivative **5.174**.

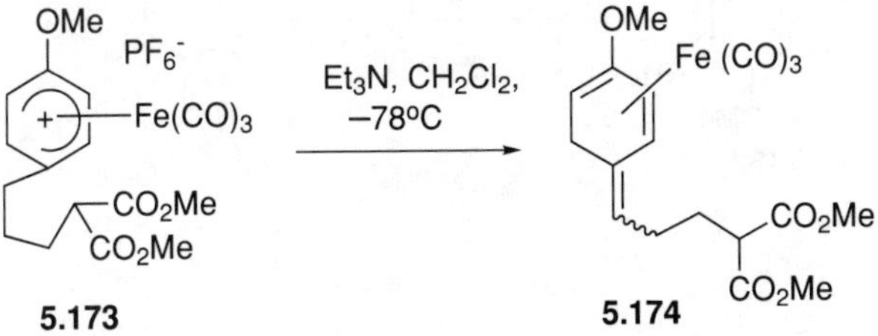

OMe ; PF_6^- ; $Fe(CO)_3$; CO_2Me ; CO_2Me — Et_3N, CH_2Cl_2, −78°C → OMe ; Fe $(CO)_3$; CO_2Me ; CO_2Me (5.70)

5.173 **5.174**

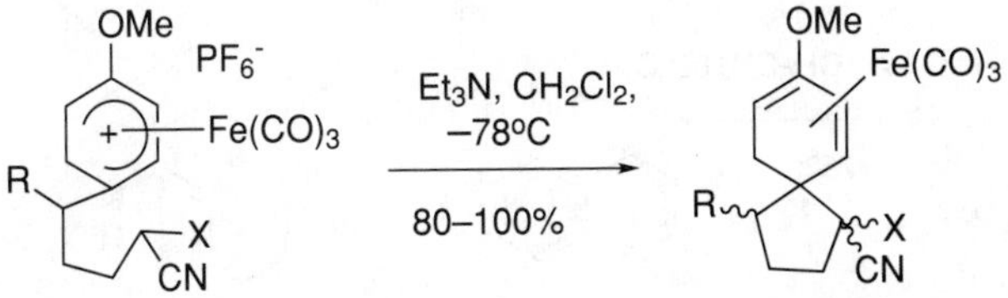

OMe ; PF_6^- ; $Fe(CO)_3$; R ; X ; CN — Et_3N, CH_2Cl_2, −78°C, 80–100% → OMe ; $Fe(CO)_3$; R ; X ; CN (5.71)

5.175 X = CN or CO_2Me, R = H
5.176 X = CN or CO_2Me, R = Me

5.177 X = CN or CO_2Me, R = H
5.178 X = CN or CO_2Me, R = Me

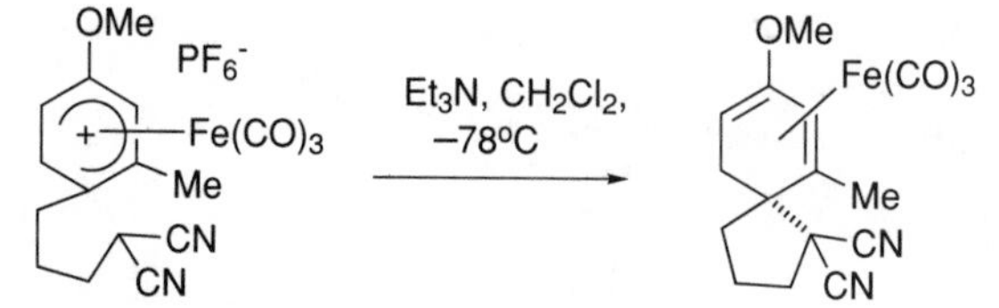

5.179 5.180 (5.72)

Some very elegant adaptations of the intramolecular nucleophile addition strategy[70] have allowed the use of an electron rich aromatic group as the nucleophile. Examples of these reactions are shown in equations 5.73–5.75, and the product **5.184** is being investigated as a potential intermediate for the synthesis of the Discorhabdin and Prianosin alkaloids[70]. An interesting but unexplained reversal in regioselectivity is obtained when the naphthylamine **5.185** is used as the nucleophile. The complexity of ring systems that might be accessible using this methodology clearly is only partially explored, and this presents a very fertile area for future research.

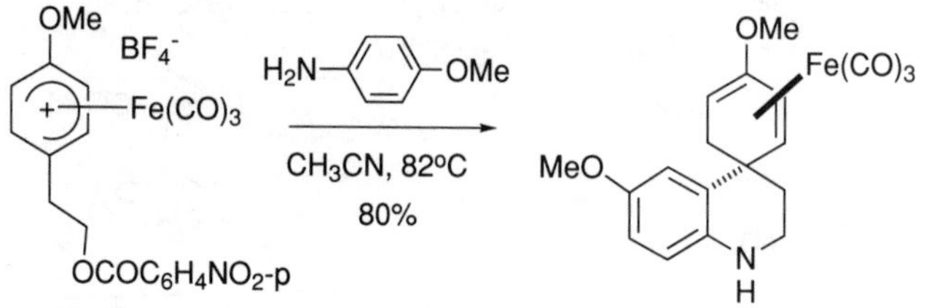

5.181 5.182 (5.73)

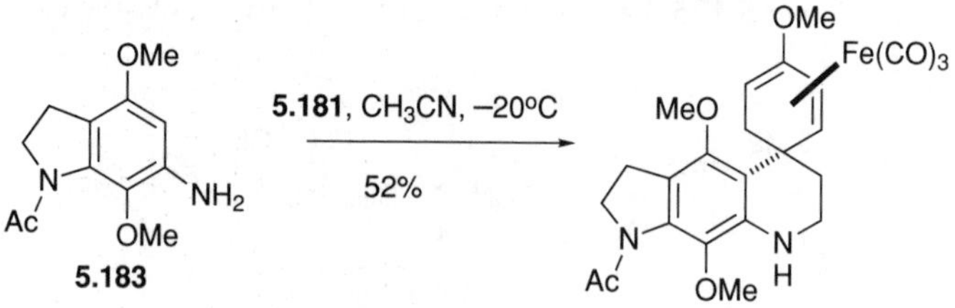

5.183 5.184 (5.74)

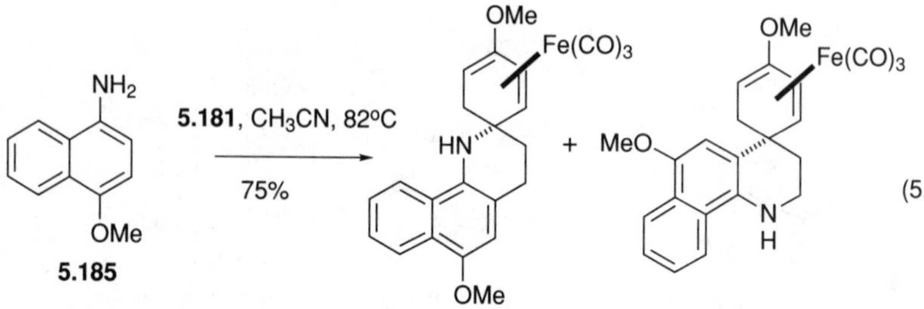

5.185 5.186 (9:1) 5.187 (5.75)

(b) Uses of synthon **5.98** *(aryl cation)*

(i) Synthesis of O*-methyljoubertiamine: aryl cation vs enone cation equivalents* *O*-Methyljoubertiamine (**5.188**), a Sceletium alkaloid, has been the target of two synthetic approaches using organoiron chemistry. The first such (formal) synthesis used the aryl cation synthon approach (**5.98,** R = OMe)[71], whereas the second used nucleophile addition to complexes of structure **5.9** (R = aryl)[72]. Both of these approaches will be described here in order to give the reader an appreciation of the flexibility for using cyclohexadienyliron complexes. The aryl cation approach is shown in Scheme 5.10, and the enone cation approach in Scheme 5.11.

Of these approaches, that shown in Scheme 5.11 is the shorter and more flexible synthesis. The use of complex **5.193** as a precursor to a potentially wide range of aryl- or alkyl- substituted dienyl systems is very exciting. It also illustrates very nicely the utility of the $Fe(CO)_3$ group in allowing transformations (H_2/Ra-Ni) that cannot be effected in its absence (in this case owing to the presence of the enone or dienol ether).

Some potentially more useful applications of the aryl cation synthon idea have been recently described[70], based on oxidative cyclization techniques developed by earlier workers (compare with equations 5.15–5.17), and on other similar

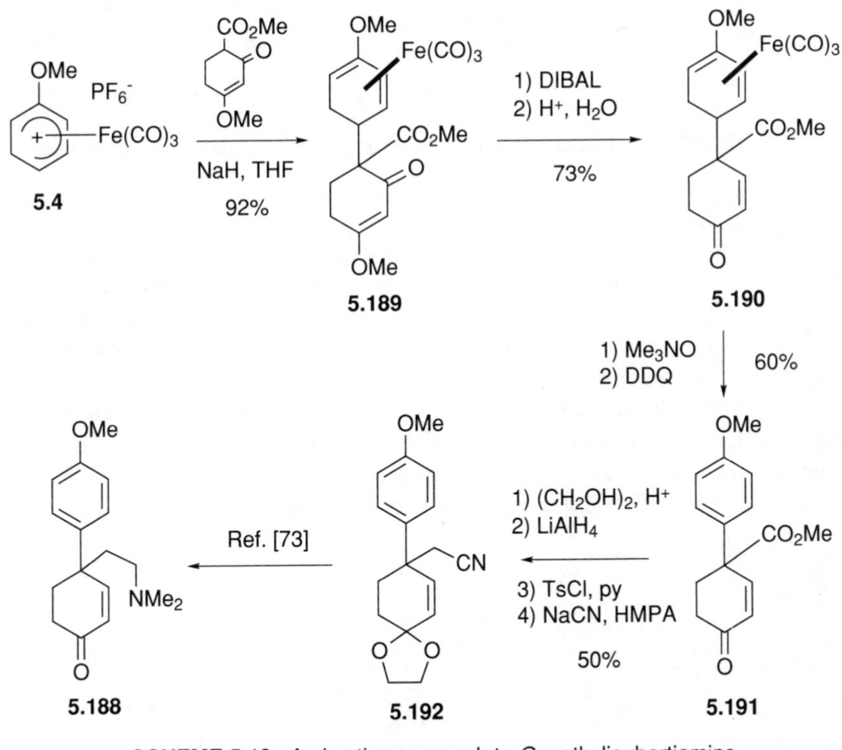

OMe PF6- Fe(CO)3 5.4 → CO2Me O OMe, NaH, THF 92% → OMe Fe(CO)3 CO2Me O OMe 5.189 → 1) DIBAL 2) H+, H2O 73% → OMe Fe(CO)3 CO2Me O 5.190

5.190 → 1) Me3NO 2) DDQ 60% → OMe CO2Me O 5.191 → 1) (CH2OH)2, H+ 2) LiAlH4 3) TsCl, py 4) NaCN, HMPA 50% → OMe CN O O 5.192 → Ref. [73] → OMe NMe2 O 5.188

SCHEME 5.10. Aryl cation approach to *O*-methyljoubertiamine.

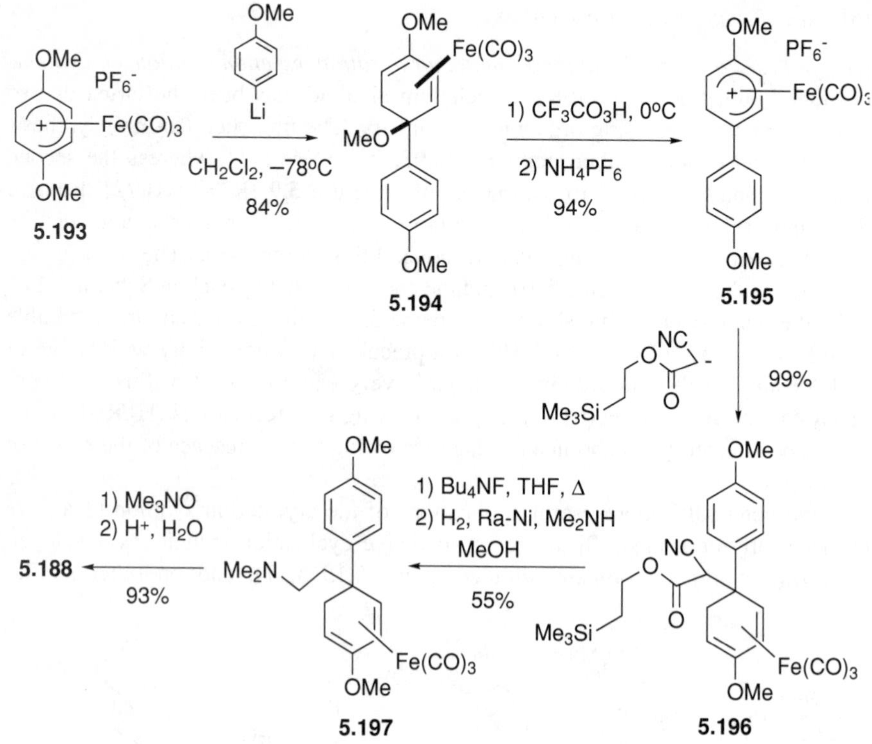

SCHEME 5.11. Cyclohexenone γ cation approach to *O*-methyljoubertiamine.

aromatization methods[74]. Initial experiments on the conversion of aminoaryl-substituted cyclohexadiene complexes by treatment with specially prepared γ-manganese dioxide gave rather poor yields of cyclization/aromatization, examples being shown in equations 5.76 and 5.77. Despite this, the method does provide a fairly direct route to deoxycarbazomycin derivatives such as **5.201**.

The overall yield of the cyclization/aromatization/demetallation sequence was improved by carrying out the conversion in a stepwise manner, as shown in equation 5.78. Chemoselective oxidation of **5.200** to the imino quinone **5.202** was accomplished in 63% yield using commercial activated manganese dioxide, and

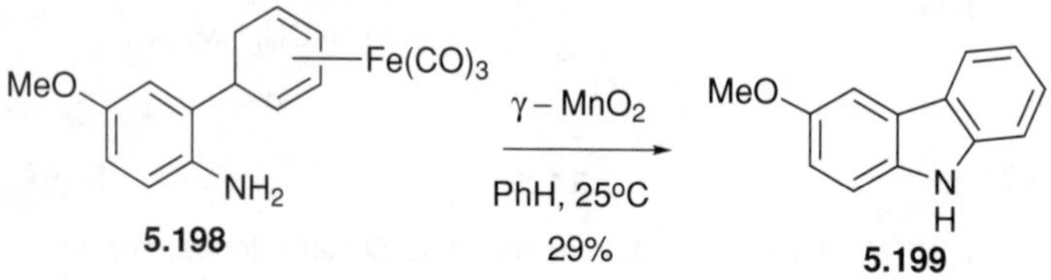

(5.76)

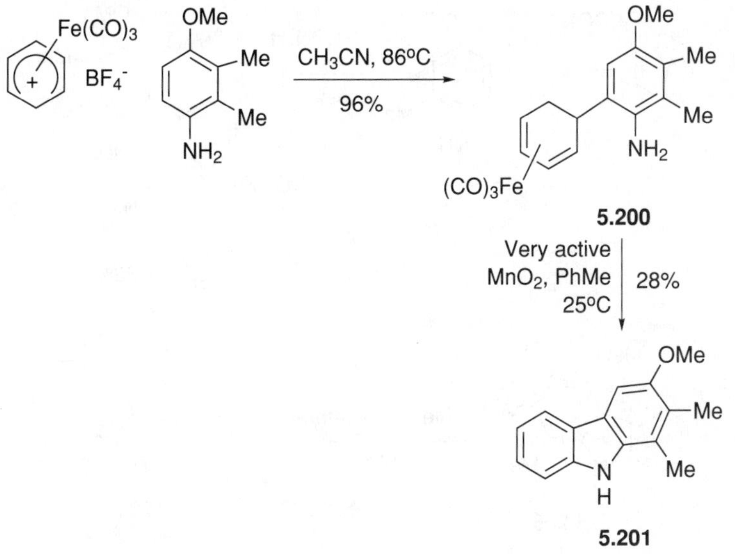

(5.77)

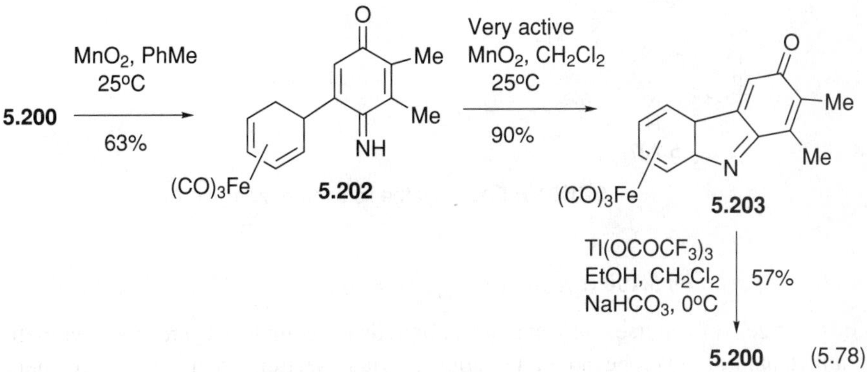

(5.78)

this was cyclized to the 4b,8a-dihydrocarbazol-3-one **5.203** in 90% yield with very active manganese dioxide. Alternatively, direct conversion of **5.200** into **5.203** could be effected in 57% yield by using thallium(III) trifluoroacetate. This method provided a higher-yielding route to the desired carbazomycin derivatives obtained by decomplexation of, e.g., **5.203**, followed by rearrangement, and a synthesis of the natural product, carbazomycin A (**5.207**) based on this approach, is outlined in Scheme 5.12. It should be noted that the cyclohexadienyliron complex is converted into a *disubstituted* aromatic ring during this sequence of reactions.

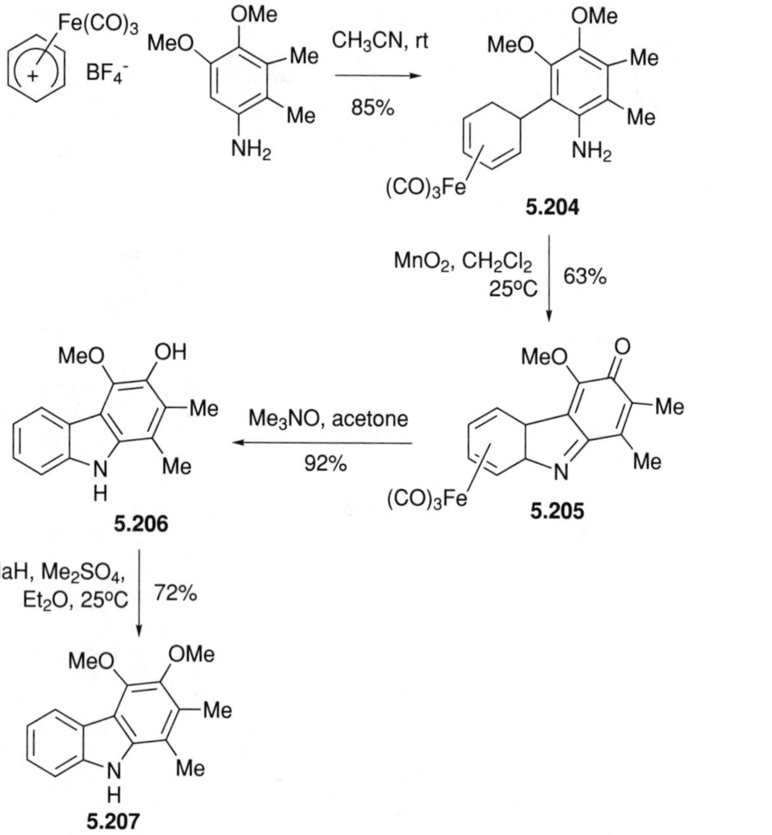

SCHEME 5.12. Synthesis of carbazomycin A.

(c) Uses of synthon **5.100** (*cyclohexadienyl cation*)

In practice, of course, *all* cyclohexadienyliron complexes are the synthetic equivalents of cyclohexadienyl cations. This section deals with synthetic applications that make use of the substituted cyclohexa*diene* functionality that is produced by nucleophile addition followed by decomplexation. The diene may be present in the final target, it may be used in subsequent Diels–Alder reactions, or it may be functionalized by selective manipulation of one of the double bonds. There are examples of all three of these applications.

(i) Synthesis of (+)- and (−)- gabaculine Using optically pure (+)-**5.11**, prepared *via* resolution as discussed earlier, Birch and co-workers[75] carried out a short synthesis of optically pure (−)-gabaculine (**5.210**) (Scheme 5.13), a compound which is of interest because of its inhibition of 4-aminobutyrate:2-oxoglutarate aminotransferase, which could be useful in the treatment of, for example,

Parkinsonism, schizophrenia, and epilepsy. The (+)- enantiomer of gabaculine can also be prepared starting from (+)-**5.11**.

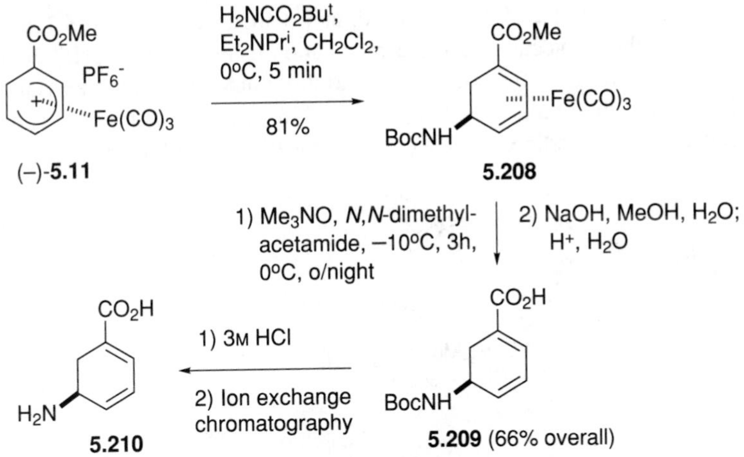

CO_2Me PF_6^- $Fe(CO)_3$
(−)-**5.11**

$H_2NCO_2Bu^t$, Et_2NPr^i, CH_2Cl_2, 0°C, 5 min
81%

CO_2Me $Fe(CO)_3$ BocNH
5.208

1) Me_3NO, *N,N*-dimethylacetamide, −10°C, 3h, 0°C, o/night
2) NaOH, MeOH, H_2O; H^+, H_2O

CO_2H BocNH
5.209 (66% overall)

1) 3M HCl
2) Ion exchange chromatography

CO_2H H_2N
5.210

SCHEME 5.13. Synthesis of (−)-gabaculine.

(ii) Synthesis of (+)- and (−)- methyl shikimate[76] A variation on the gabaculine synthesis, in which hydroxide is added to (+)-**5.11** provided access to the hydroxy-substituted diene complex **5.211**, (Scheme 5.14) which was converted into the demetallated silyl ether derivative **5.212**. Selective osmylation of the more reactive double bond in this intermediate, followed by desilylation furnished (−)-methyl shikimate (**5.213**) (shikimic acid is an important intermediate in biosynthesis).

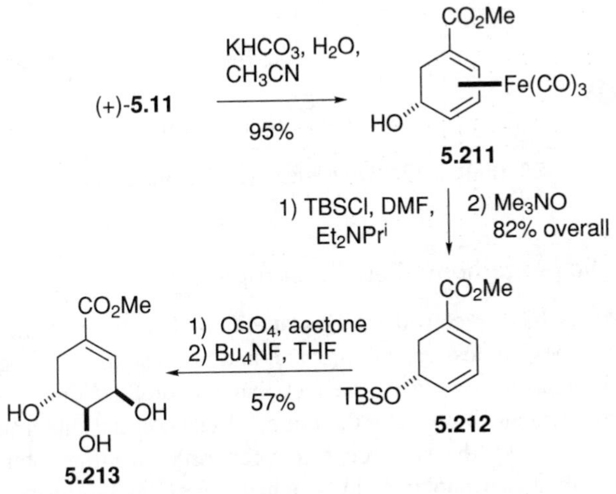

(+)-**5.11**

$KHCO_3$, H_2O, CH_3CN
95%

CO_2Me $Fe(CO)_3$ HO
5.211

1) TBSCl, DMF, Et_2NPr^i
2) Me_3NO
82% overall

CO_2Me TBSO
5.212

1) OsO_4, acetone
2) Bu_4NF, THF
57%

CO_2Me HO OH OH
5.213

SCHEME 5.14. Synthesis of (−)-methyl shikimate.

(iii) Synthesis of dihydrocannivonine[77] This synthesis takes advantage of the moderated reactivity of the dicarbonyl(triphenylphosphine)cyclohexadienyl complex **5.65** in securing high-yield alkylation with Grignard reagents (Scheme 5.15). In this manner, intermediate **5.214** was prepared, and demetallated to give **5.215** (note the loss of THP ether protection during demetallation using $CuCl_2$/EtOH). Oxidation of the alcohol, followed by treatment of the resulting aldehyde with methylammonium chloride gave the iminium derivative **5.216** which underwent spontaneous intramolecular cycloaddition to give (±)-dihydrocannivonine (**5.218**).

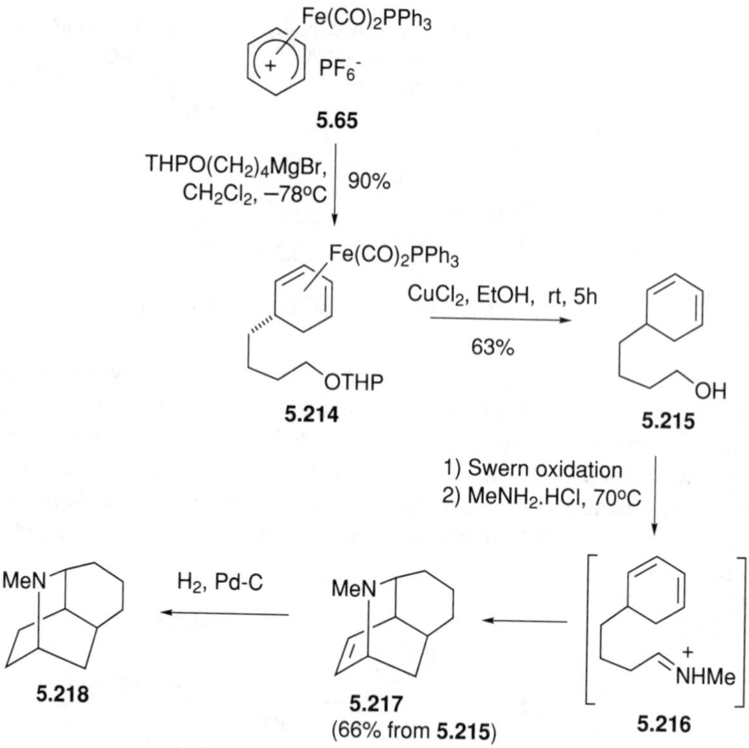

$Fe(CO)_2PPh_3$ PF_6^-

5.65

$THPO(CH_2)_4MgBr$, CH_2Cl_2, −78°C 90%

$Fe(CO)_2PPh_3$ OTHP **5.214**

$CuCl_2$, EtOH, rt, 5h 63%

OH **5.215**

1) Swern oxidation
2) $MeNH_2$.HCl, 70°C

$\overset{+}{N}HMe$ **5.216**

MeN **5.217** (66% from **5.215**)

H_2, Pd-C

MeN **5.218**

SCHEME 5.15. Synthesis of (±)-dihydrocannivonine.

5.1.4 Acyclic pentadienyl-$Fe(CO)_3$ complexes

These complexes have received far less attention than their cyclic counterparts, largely because they are less easy to prepare, and because the nucleophilic addition is not as well-behaved, owing to the fact that the product diene complexes are prone to rearrangement. Moreover, there are fewer potential direct applications of these complexes in synthesis, except in preparing substrates for the type of chemistry described in Chapter 4, in which the $Fe(CO)_3$ group is used as diene protection and as a stereocontroller.

Only in special cases can hydride abstraction be used as a method for converting acyclic diene complexes into the dienyl cations. The reason is that a *cisoid* geometry, as shown in structure **5.219,** is required for this reaction to proceed. Reaction of, for example, (*Z*)-1,3-pentadiene (**5.220**) with iron pentacarbonyl does not give **5.219**, but the product (**5.221**) of isomerization and complexation. Reaction of **5.221** with trityl cation gives no dienyl product. In certain cases it may be possible to prepare a diene complex in which an appropriate geometry exists, e.g., **5.222**, and then hydride abstraction proceeds cleanly to give **5.223**[78].

Fe(CO)3 / Me **5.219** ← (Fe(CO)5, crossed arrow) — Me **5.220** — Fe(CO)5 → Fe(CO)3 / Me **5.221** (5.79)

Me / Me / Me — Fe2(CO)9, acetone → Me Fe(CO)3 / Me / Me **5.222** — Ph3CBF4 → Me Fe(CO)3 / + / Me / BF4- **5.223** (5.80)

The most convenient method for preparing these compounds is by acid treatment of pentadienol-Fe(CO)$_3$ complexes, examples being shown in equations 5.81 and 5.82. The complexation of dienols and dienals proceeds quite satisfactorily, and good yields are obtained.

Fe(CO)3 / R / CH2OH — HClO4 or HPF6 → Fe(CO)3 / + / X- / R (5.81)

5.224 R = H
5.225 R = Me

5.226 R = H
5.227 R = Me

Me / CHO — 1) MeMgI 2) Fe(CO)5 → Fe(CO)3 / Me / CH(OH)Me **5.228** — HBF4 → Fe(CO)3 / + / BF4- / Me / Me **5.229** (5.82)

Reactions of these complexes with nucleophiles usually occur at the dienyl terminus to give substituted diene complexes, the only exception being that shown in equation 5.88[79]. If the nucleophile is fairly reactive, the cisoid geometry is maintained, but in some cases rearrangement can occur to give the more stable *trans* geometry. This is especially true if heat is required, or if the product complexes are treated with acid. Some examples of nucleophile addition are shown in equation 5.83–5.91[80–83].

Fe(CO)3 BF4- R Me H2O → Fe(CO)3 R CH(OH)Me (5.83)

5.230 R = Ph or Me **5.231** R = Ph or Me

Fe(CO)3 BF4- RNH2 → Fe(CO)3 CH2NHR + Fe(CO)3 R N Fe(CO)3 (5.84)

5.226 **5.232** (*E*/*Z* mixture) **5.233** (*E*/*Z* mixture)

Me Fe(CO)3 Me BF4- NaCH(CO2Me)2 → Me Fe(CO)3 Me CH(CO2Me)2 (5.85)

5.223 **5.234**

Fe(CO)3 Me PF6- NaCH(CO2Me)2 THF 53% → Fe(CO)3 Me CH(CO2Me)2 + Me Fe(CO)3 CH(CO2Me)2 (5.86)

5.227 **5.235** (1:1.8) **5.236**

Fe(CO)3 Ph PF6- LiCH(CO2Me)2 THF 50% → Ph Fe(CO)3 CH(CO2Me)2 (5.87)

5.237 **5.238**

Fe(CO)3 EtO2C PF6- LiCH(CO2Me)2 → (MeO2C)CH EtO2C (CO)3Fe (5.88)

5.239 **5.240**

$Fe(CO)_3$ BF_4^- —R_2Cd→ $Fe(CO)_3$ R (5.89)

R $Fe(CO)_3$ PF_6^- —$(R'C{\equiv}C)_3CuLi_2$, Et_2O, THF, −65°C→ R $Fe(CO)_3$ R′ **5.241** (5.90)

R = Me, Ph or CO_2Me

MeO_2C $Fe(CO)_3$ PF_6^- **5.239** —$R'C{\equiv}CSiMe_3$, KI, KF, CH_2Cl_2, Δ→ MeO_2C $Fe(CO)_3$ R′ (5.91)

Possible areas of application of this chemistry might include synthetic approaches to the leukotrienes[84,85], and in this context it may be noted that selective hydrogenation of the alkyne in complexes such as **5.242** can be accomplished in high yield (equation 5.92)[86]. The unusual C-2 addition product **5.240** can be converted into vinylcyclopropanes by treatment with ceric ammonium nitrate (equation 5.93), but very little work has been done to establish the generality of this method for cyclopropane synthesis[79].

MeO_2C $Fe(CO)_3$ C_5H_{11} **5.241** —H_2, Lindlar cat., 91%→ MeO_2C $Fe(CO)_3$ C_5H_{11} (5.92)

$(MeO_2C)CH$ EtO_2C $(CO)_3Fe$ **5.240** —Ce(IV), MeOH, 76%→ $CH(CO_2Me)_2$ EtO_2C (5.93)

5.1.5 Cycloheptadienyliron complexes

These complexes had been studied in the 1970s[87–89] and some of the reactivity patterns were delineated, although there were no attempts to apply them in synthesis until some ten years later. Cycloheptadienyl-$Fe(CO)_3$ **5.243** is readily prepared by the usual hydride abstraction method, or by protonation of the η^4-triene complex **5.244.** The main problem with this complex is that nucleophile addition reactions are somewhat unreliable, often giving mixtures of regioisomers, as well as low yields. Some typical results are shown in equations 5.95–5.99.

$Fe(CO)_3$ **5.242** $\xrightarrow{Ph_3CPF_6}$ $Fe(CO)_3$ PF_6^- **5.243** $\xleftarrow{HPF_6}$ $Fe(CO)_3$ **5.244** (5.94)

5.243 $\xrightarrow[62\%]{NaCH(CO_2Et)_2}$ $Fe(CO)_3$ $CH(CO_2Et)_2$ **5.245** (5.95)

5.243 $\xrightarrow[61\%]{NaOEt}$ $Fe(CO)_3$ OEt **5.246** (5.96)

5.243 $\xrightarrow[90\%]{NaBH_4}$ $Fe(CO)_3$ **5.242** + $(CO)_3Fe$ **5.247** (33:66) (5.97)

5.243 $\xrightarrow[53\%]{NaCN}$ $Fe(CO)_3$ CN **5.248** + NC $(CO)_3Fe$ **5.249** (28:72) (5.98)

5.243 $\xrightarrow{Me_2CuLi}$ $Fe(CO)_3$ Me **5.250** (*ca* 15%) + $Fe(CO)_3$ $Fe(CO)_3$ **5.251** (*ca* 15%) (5.99)

Work on cyclohexadienyliron complexes had suggested that sequential nucleophilic additions could be achieved in a completely stereocontrolled manner, using the directing power of the $Fe(CO)_3$ group[16,17]. For example, complex **5.44** *(vide supra)*, which was obtained via malonate addition to complex **5.9**, could be converted into the dienyl complex **5.46**, admittedly using a rather tortuous route, and this complex reacted with malonate anion or dimethylcopperlithium to give complexes **5.252** or **5.253** (equation 5.100). Thus, 1,2-stereocontrol can be effected. The seven-membered ring in **5.243** provides an opportunity for 1,3- stereocontrol, but it was clear that a little work was needed to obtain good control during the nucleophile additions. Particularly disturbing, but informative, was the result of cuprate addition (equation 5.99). The dimer **5.251**, is the result of an electron transfer to **5.243**, and the observation of this pathway suggested a means to control the reactivity of the dienylmetal system.

OMe, $Fe(CO)_3$, PF_6^-, Me, OAc — **5.46** $\xrightarrow[\text{or } Me_2CuLi]{NaCH(CO_2Me)_2}$ OMe, $Fe(CO)_3$, R, Me, OAc (5.100)

5.252 R = $CH(CO_2Me)_2$
5.253 R = Me

Replacement of one CO ligand on **5.243** by a poorer π-acceptor, e.g., triphenylphosphine, should raise the reduction potential of the complex, thereby minimizing electron transfer reactions. Furthermore, it was noted that during the reaction of **5.243** with Me_2CuLi, a large amount of polar, uncharacterized material was formed, and this was thought to be the result of nucleophile addition to a carbonyl ligand to give an unstable metal acyl species. The introduction of a phosphine ligand would also eliminate this pathway, by deactivating the remaining carbonyl ligands. Treatment of complex **5.242** with triphenylphosphine in boiling di-n-butyl ether gave **5.254**, the desired ligand exchange product, but in variable yields (50–70%). Much better results were obtained with triphenylphosphite, giving **5.255** in greater than 95% yield. Since the dienyl complexes **5.256** and **5.257** showed virtually identical reactivity towards nucleophiles, it was decided to use the latter, which could be readily prepared on a scale of 250 g (or more, if desired). As with the cyclohexadienyl complexes, these materials have an excellent shelf life and are very easy to use.

$Fe(CO)_3$ — **5.242** $\xrightarrow{PR_3,\ \Delta}$ $Fe(CO)_2PR_3$ $\xrightarrow[98\text{–}100\%]{Ph_3CPF_6}$ $Fe(CO)_2PR_3$, PF_6^- (5.101)

5.254 R = Ph (50–70%)
5.255 R = OPh (>95%)

5.256 R = Ph
5.257 R = OPh

Reactions of complex **5.257** with carbon nucleophiles are exceptionally well behaved[90]. Depending on the nature of the nucleophile, *either* diene complexes *or* enediyl complexes (see structures **5.247** and **5.249**) can be obtained in very high yield (equations 5.102–5.106). Mixtures are not produced. It may be noted that cyclohexadienyliron complexes never give products of addition at C-2 of the dienyl; the reasons for this difference between the two ring sizes are not well understood. One possible drawback is the cost of cycloheptadiene (Aldrich catalog 1992, $119.70 for 10 g), however, this substance is readily prepared by lithium/ammonia reduction of cycloheptatriene ($80.60 l^{-1}).

$Fe(CO)_2P(OPh)_3$ + PF_6^- **5.257** —[$NaCH(CO_2Me)_2$, 99%]→ $Fe(CO)_2P(OPh)_3$, $CH(CO_2Me)_2$ **5.258** (5.102)

5.257 —[NaOMe, 95%]→ $Fe(CO)_2P(OPh)_3$, OMe **5.259** (5.103)

5.257 —[Me_2CuLi, 97%]→ $Fe(CO)_2P(OPh)_3$, Me **5.260** (5.104)

5.257 —[MeLi, 94%]→ Me, $(CO)_2[(PhO)_3P]Fe$ **5.261** (5.105)

5.256 —[$NaBH_4$ or NaCN]→ R, $(CO)_2[Ph_3P]Fe$ (5.106)

5.262 R = H
5.263 R = CN

In contrast to the products of nucleophile addition to cyclohexadienyliron complexes, which are resistant to hydride abstraction by trityl cation, the seven-membered ring derivatives undergo this reaction smoothly. For example, **5.260** is converted into **5.264** in virtually quantitative yield. Nucleophile additions to **5.264** proceed in excellent yield with complete stereo- and regiocontrol (*anti* to the metal at the less hindered dienyl terminus), examples being given in equation 5.107.

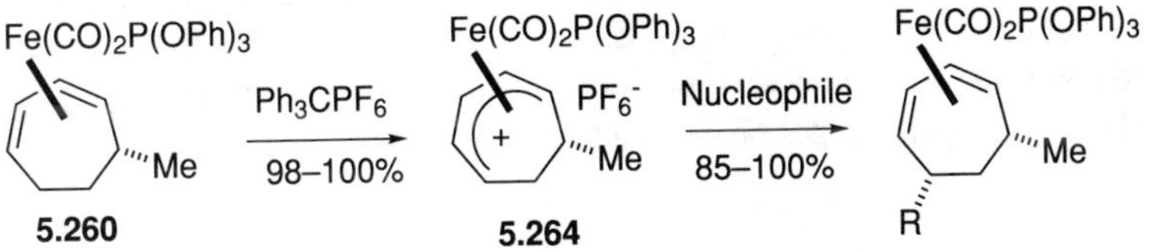

$Fe(CO)_2P(OPh)_3$ **5.260** → (Ph_3CPF_6, 98–100%) → $Fe(CO)_2P(OPh)_3$ PF_6^- **5.264** → (Nucleophile, 85–100%) → $Fe(CO)_2P(OPh)_3$ (5.107)

5.265 R = Me (99%)
5.266 R = $CH(CO_2Me)_2$ (95–98%)
5.267 R = $CH_2CH{=}CH_2$

The *syn* 1,3- relationship of methyl substituents in **5.265** and methyl/malonate groups in **5.266** is appropriate for applications of this methodology in synthetic approaches to a number of important macrolide antibiotics, provided the metal can be removed, the diene can be functionalized in a controlled manner, and the ring can be cleaved to give acyclic molecules. Such operations have in fact been carried out successfully, and subunits for several macrolides have been prepared efficiently. The approach will be illustrated here with the synthesis of optically pure (+)-Prelog–Djerassi lactone (**5.273**, Scheme 5.16), a synthetic building block and degradation product of macrolides such as methymycin[91].

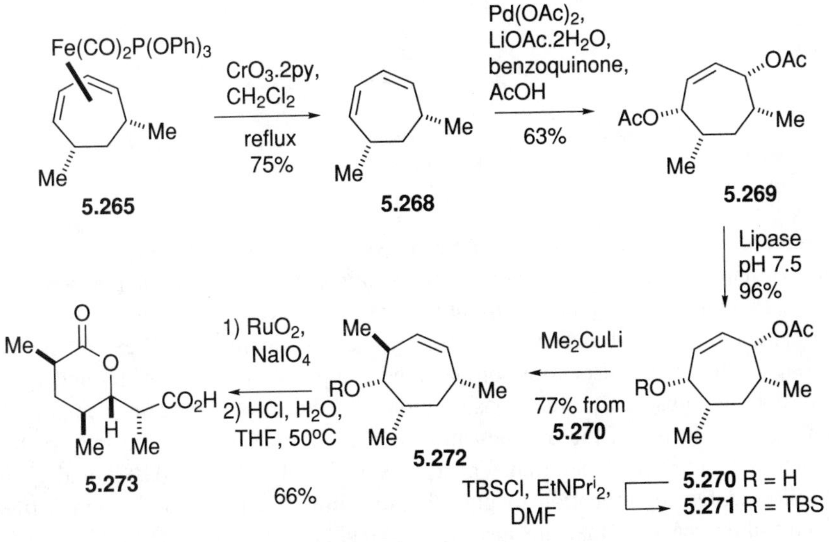

5.265 → (CrO_3.2py, CH_2Cl_2, reflux, 75%) → **5.268** → ($Pd(OAc)_2$, LiOAc.$2H_2O$, benzoquinone, AcOH, 63%) → **5.269** → (Lipase, pH 7.5, 96%) → **5.270** R = H; TBSCl, $EtNPr^i_2$, DMF → **5.271** R = TBS; **5.271** → (Me_2CuLi, 77% from **5.270**) → **5.272** → (1) RuO_2, $NaIO_4$; 2) HCl, H_2O, THF, 50°C; 66%) → **5.273**

SCHEME 5.16. Synthesis of (+)-Prelog–Djerassi lactone.

From Scheme 5.16 it can be seen that the diene moiety of **5.268** can be functionalized with excellent stereocontrol. Bäckvall's diacetoxylation method[92] effects conversion into the *meso* diacetate **5.269**, obtained as a single diastereomer, and enzymatic hydrolysis allows asymmetrization to give optically pure hydroxy acetate **5.270.** Protection, followed by *anti* displacement of the allylic acetate by using dimethylcuprate, sets in place the remaining methyl group with complete stereocontrol, and ring scission followed by deprotection and lactonization affords the target molecule. The synthesis is quite short (10 steps from **5.257**, *ca* 22% overall yield) and does not involve separation of diastereomeric mixtures (except that the enzyme hydrolysis cannot be carried to completion).

5.2. FERROCENE DERIVATIVES

5.2.1 Background

Ferrocene (**5.274**) was first prepared[93] by the reaction of cyclopentadienyl-magnesium bromide with $FeCl_3$, and is obtained as an orange crystalline solid, m.p. 174°C. This compound occupies a princely position in organometallic chemistry, since its discovery led to a re-birth of this area of study and the subsequent explosive growth of research over a very wide range of transition metal complexes. Ferrocene and its derivatives find many applications in materials science and in general synthetic chemistry. Their chemistry is reviewed annually[94], and is dealt with on a regular basis in Gmelin's Handbüch[95]. A number of methods are available for the preparation of ferrocene, but since it is commercially available and inexpensive, these are hardly worth pursuing.

- M+ FeCl$_3$ or FeCl$_2$ → Fe (5.108)

M = Li, Na, K, MgX, etc. **5.274**

Ferrocene is remarkably stable, and can be subjected to a wide range of chemical transformations. These are important in giving access to substituted ferrocenes, some of which have become important in organic synthesis. We shall briefly review the salient features of ferrocene chemistry, bearing in mind that examples will appear in the later discussion on synthetic applications. Ferrocene undergoes electrophilic substitution reactions, similar to those of aromatic compounds, although the details of the mechanism are different, the electrophilic reagent first adding to the metal and then transferring to one of the Cp (Cp = C_5H_5) ligands. This is followed by deprotonation to give the substituted ferrocene (equation 5.109). Electrophilic reagents that are also strong oxidizing agents (e.g., NO_2^+) usually accept an electron from the metal to give the ferricinium cation, although this can be reduced back to ferrocene itself using common reducing agents such as tin(II) chloride, sodium hydrogen sulfite or ascorbic acid.

Fe —(E^+)⇌ [Fe—E]$^+$ ⇌ [Fe, H, E]$^+$ —($-H^+$)→ Fe–E (5.109)

Ferrocene is *very* reactive toward electrophiles and undergoes acetylation under Friedel–Crafts conditions approximately 3 million times faster than benzene! The acetyl group deactivates the molecule and the monoacetyl derivative **5.275** is readily isolated. This compound can be further acetylated, however, to give the *heteroannularly* disubstituted compound **5.276** (approx. 200 times faster than benzene itself). Vilsmeier formylation is also highly selective (equation 5.111). Complexes **5.275** and **5.277** undergo normal ketone and aldehyde reactions, such as reduction, oxidation, aldol condensation, Wittig olefination, etc. Some of these are summarized in equations 5.112–5.119.

5.274 —(CH_3COCl, $AlCl_3$, CH_2Cl_2, 0°C)→ **5.275** (COMe) —(CH_3COCl, $AlCl_3$, CH_2Cl_2; Slower)→ **5.276** (COMe, COMe) (5.110)

5.274 —(1) PhN(Me)CHO, $POCl_3$; 2) H_2O)→ **5.277** (CHO) (5.111)

5.275 (COMe) —(RCHO, base)→ **5.278** (COCH=CHR) (5.112)

5.275 —(RCO_2Et, KNH_2)→ **5.279** ($COCH_2COR$) (5.113)

5.277 (CHO) —(KOH, EtOH; Cannizzaro reaction)→ **5.280** (CH_2OH) + **5.281** (CO_2H) (5.114)

$$\textbf{5.277} \xrightarrow[\text{Knoevenagel}]{CH_2(CO_2H)_2,\ \text{py., piperidine}} \textbf{5.282}\ (\text{Fc–CH=CHCO}_2\text{H}) \tag{5.115}$$

$$\textbf{5.277} \xrightarrow{Ph_3P{=}CHR} \textbf{5.283}\ (\text{Fc–CH=CHR}) \tag{5.116}$$

$$\textbf{5.277} \xrightarrow[2)\ PCl_5]{1)\ NH_2OH} \textbf{5.284}\ (\text{Fc–CN}) \tag{5.117}$$

$$\textbf{5.280} \xrightarrow{MnO_2} \textbf{5.277} \tag{5.118}$$

$$\textbf{5.284} \xrightarrow{SnCl_2,\ HCl} \textbf{5.277} \tag{5.119}$$

Metallation of ferrocene allows access to substituted derivatives that are not readily available by direct electrophilic substitution. Monolithiation, using excess Bu^nLi in diethyl ether, proceeds in only 25% yield, unreacted ferrocene accounting for the remaining material. Because of the potential importance of monolithioferrocene, more efficient methods for its preparation have been developed, *via* metal–halogen exchange (Bu^nLi + bromoferrocene)[96] or transmetallation (Bu^nLi + chloromercuriferrocene)[97,98]. Recently, better reaction conditions for direct monolithiation of ferrocene (equation 5.120) have been developed[99]. The metallated derivative **5.285** undergoes reactions with electrophiles to give substituted ferrocenes, as expected (equations 5.121–5.127). As we shall see later, metallation of substituted ferrocenes provides access to important chiral ferrocenylphosphines that can be used as ligands for asymmetric catalytic reactions.

$$\textbf{5.274}\ (\text{Fe}) \xrightarrow[\text{Schlenk tube}]{Bu^tLi,\ THF,\ 0^oC,\ 15\ \text{min}} \textbf{5.285}\ (\text{Fc–Li}) \tag{5.120}$$

$$\textbf{5.285}\ (\text{Fc–Li}) \xrightarrow[2)\ H^+]{1)\ CO_2} \textbf{5.281}\ (\text{Fc–CO}_2\text{H}) \tag{5.121}$$

5.285 —N_2O_4→ NO_2 Fe **5.286** —Fe, HCl→ NH_2 Fe **5.287** (5.122)

5.285 —NH_2OMe→ **5.287** (5.123)

5.285 —1) $B(OR)_3$ 2) hydrolysis→ $B(OH)_2$ Fe **5.288** (5.124)

5.285 —CH_3COCH_3, 85%→ Me Me OH Fe **5.289** (5.125)

5.285 —O, 85%→ Fe OH **5.290** (5.126)

5.285 —$HgCl_2$→ HgCl Fe **5.291** —X_2, X = Cl or Br→ X Fe **5.292** (5.127)

An important property of the ferrocenyl group, which allows useful transformations of the chiral derivatives that we shall discuss later, is its ability to stabilize carbocations *alpha* to the cyclopentadienyl ring, leading to neighboring group participation (NGP) during solvolysis and related reactions. This is a result of interaction of the developing positive charge in the transition state with filled iron d orbitals, and leads to retention of configuration during certain transformations (*vide infra*). Examples of this behavior are shown in equations 5.128–5.130; particularly striking is the observation that *exo* acetate **5.296** undergoes solvolysis 2500 times faster than the *endo* acetate **5.298**, indicative of the aforesaid NGP.

Fe **5.293** —AcOH→ [Fe $\overset{+}{C}HCH_3$ **5.294**] —AcO^-→ Fe OAc **5.295** (5.128)

Fe OAc H **5.296** —H_2O, acetone→ Fe OH H **5.297** (5.129)

Fe H OAc **5.298** —H_2O, acetone→ **5.297** (5.130)

5.2.2 Applications of chiral ferrocenylphosphine derivatives in organic synthesis

This is probably one of the most important developments in ferrocene chemistry as applied to organic synthesis. Homoannularly disubstituted ferrocenes, in which the substituents are non-equivalent, show planar chirality. In addition, chiral molecules can be prepared in which the substituent itself contains an asymmetric center. Both types of chirality play an important role in the development of ferrocenyl derivatives that can be used as chiral ligands for asymmetric catalysis. It is important to note that the iron itself does not become involved in the catalytic cycle, but instead provides the framework for some very useful ligands. An important development in this respect was the preparation of optically pure α-ferrocenylethyldimethylamine (**5.300**) by resolution using (*R*)-(+)-tartaric acid[100]. The racemic complex **5.299** is itself readily prepared either by phosgenation of α-ferrocenylethanol followed by treatment with dimethylamine, or by reaction of ferrocenylethyldimethylaminoacetonitrile (**5.299**) with methylmagnesium iodide (equation 5.131)[101].

Fe CHO **5.277** —Me_2NH; NaCN, $NaHSO_4$, 90%→ Fe CN NMe_2 **5.299** —MeMgI→ Fe Me NMe_2 **5.300** (5.131)

Lithiation of **5.300** occurs with pronounced regio- and diastereoselectivity, allowing the construction of disubstituted ferrocenes having planar chirality with excellent stereocontrol (equation 5.132)[100]. This has led to methods for preparing optically pure ferrocenylphosphine derivatives which have been extensively investigated as chiral ligands for asymmetric catalysis, some of which will now be discussed.

(*R*)-**5.300** → 1) Bu^nLi; 2) Electrophile → (*R*,*S*)-**5.301** (96%) (*R*,*R*)-**5.301** (4%)
R = $SiMe_3$, or CH_2OH, or Ph_2COH (5.132)

A review of this area appeared in 1982[102], describing early progress in asymmetric catalysis. There have been several advances reported in the years subsequent to this report, and the types of transformation that can be effected will be reviewed here. Reaction of (*R*) - **5.300** with Bu^nLi in diethyl ether, followed by diphenylphosphinyl chloride, gives mainly (*R*)-*N*,*N*-dimethyl-1-[(*S*)-2-(diphenylphosphino)ferrocenyl]ethylamine [(*R*)-(*S*)-PPFA, (**5.302**)], which is obtained diastereomerically pure by simple recrystallization. The absolute configuration of **5.302** has been confirmed by X-ray crystallography. The minor diastereomer obtained from this reaction (*R*,*R*-**5.302**) can be obtained in quantity by lithiation of the trimethylsilyl derivative (*R*,*S*)-**5.301**, followed by reaction with Ph_2PCl and desilylation (TMS acts as a blocking group during the *ortho*-lithiation). Treatment of (R,S)-**5.302** with Bu^nLi in *N*,*N*,*N′*,*N′*-tetramethylethylenediamine (TMEDA), followed by Ph_2PCl results in the formation of the heteroannular bisphosphine derivative **5.303** [(*R*)-(*S*)-BPPFA].

(*R*)-**5.300** → 1) Bu^nLi; 2) Ph_2PCl; crystallize → (*R*,*S*)-**5.302** → BuLi, TMEDA; Ph_2PCl → (*R*,*S*)-**5.303** (5.133)

(*S*,*R*)-**5.302** or **5.303** → Ac_2O → (*S*,*R*)-**5.304** Y = H; (*S*,*R*)-**5.305** Y = PPh_2 → NaOMe, MeOH, Δ → (*S*,*R*)-**5.306** Y = H; (*S*,*R*)-**5.307** Y = PPh_2 (5.134)

By making use of the NGP effects mentioned in the preceding section, the dimethylaminoethyl side chain of **5.302** and **5.303** can be manipulated with excellent stereocontrol to give a variety of functionalized molecules, some examples being shown in equation 5.134.

The related optically active ferrocenylphosphine, 1-[(dimethylamino)methyl]-2-(diphenylphosphino)ferrocene (**5.308**), which has only planar chirality, has been prepared by resolution of the phosphine sulfide dibenzoyltartaric acid salt[103], and it is also possible to prepare the ethyl derivative **5.309**.

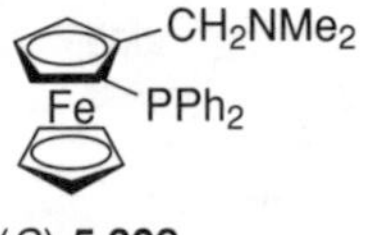

CH_2NMe_2, Fe, PPh_2 — (*S*)-**5.308**

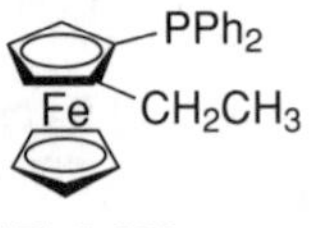

PPh_2, Fe, CH_2CH_3 — (*R*)-**5.309**

Among the first reactions to be investigated, using these chiral ferrocenylphosphines as ligands, were asymmetric cross-coupling reactions between Grignard reagents and alkyl halides, catalyzed by nickel or palladium complexes. Since the Grignard reagent undergoes rapid racemization under the reaction conditions, the asymmetric induction is a result of kinetic resolution coupled with a rapid pre-equilibration that ultimately leads to optically active product from *all* of the Grignard reagent. Two examples are shown in equations 5.135 and 5.136, together with results of asymmetric induction produced by different ferrocenylphosphines. It can be seen that the heteroatom substituent (N or O) is essential for asymmetric induction, indicating that coordination of the Grignard reagent to this substituent is involved. On the other hand, chirality in this side chain is not required (see result using **5.308**) and the most important feature of these ligands is their planar chirality. A marked improvement in asymmetric inductions is observed when an alkylzinc is used instead of the Grignard reagent (equation 5.136).

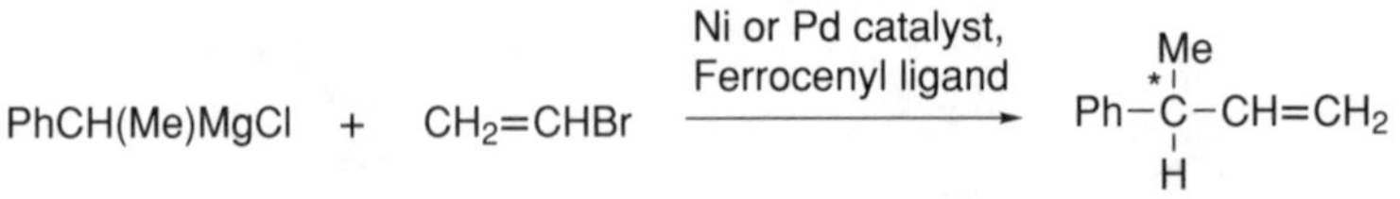

$$\mathrm{PhCH(Me)MgCl} + \mathrm{CH_2{=}CHBr} \xrightarrow[\text{Ferrocenyl ligand}]{\text{Ni or Pd catalyst,}} \mathrm{Ph{-}\overset{*}{C}H(Me){-}CH{=}CH_2} \quad (5.135)$$

5.310

Ligand	%e.e.	Abs. config^n
(*S*,*R*)-**5.302**	63	*R*
(*S*,*R*)-**5.306**	57	*R*
(*S*)-**5.308**	65	*S*
(*R*)-**5.309**	5	*S*

$$\mathrm{PhCH(Me)ZnX} + \mathrm{CH_2{=}CHBr} \xrightarrow[\text{THF}]{\mathrm{PdCl_2},\ (R,S)\text{-}\mathbf{5.302},} (S)\text{-}\mathbf{5.310}\ (85\%\ \text{e.e}) \quad (5.136)$$

Asymmetric hydrogenation of unsaturated organic molecules is an important method for preparing enantiomerically enriched compounds. Ferrocenyl bisphosphine ligands have shown promise as ligands for these transformations, and examples are given in equations 5.137–5.140.

The high asymmetric induction that is observed using the ligand (*R*, *S*)-**5.311** is possibly due to a hydrogen bonding effect between the carbonyl group of the substrate and the OH of the ligand. In a number of cases, such as equations 5.139 and 5.140 the ferrocenyldiphosphine ligands are superior to the more traditional DIOP- or DIPAMP rhodium systems.

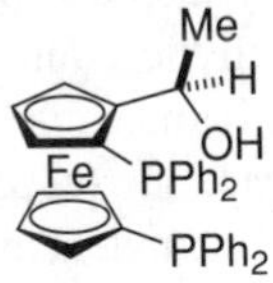

Me H OH Fe PPh_2 PPh_2

(*R*,*S*)-**5.311**

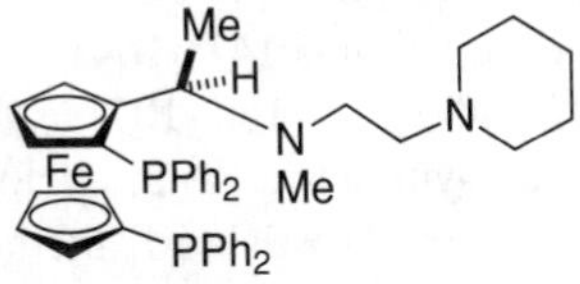

Me H N Me N Fe PPh_2 PPh_2

(*R*,*S*)-**5.312**

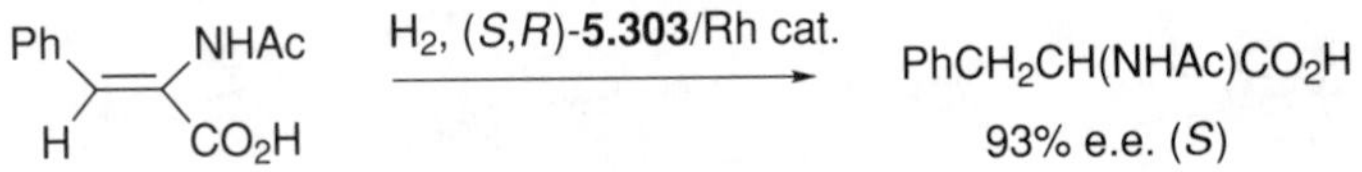

$$\text{Ph(H)C=C(NHAc)CO}_2\text{H} \xrightarrow{\text{H}_2,\ (S,R)\text{-}\mathbf{5.303}\text{/Rh cat.}} \text{PhCH}_2\text{CH(NHAc)CO}_2\text{H} \quad 93\%\ \text{e.e.}\ (S) \tag{5.137}$$

$$\text{MeCOR} \xrightarrow{\text{H}_2,\ (R,S)\text{-}\mathbf{5.311}\text{/Rh cat.}} \text{MeCH(OH)R} \tag{5.138}$$

R = Ph 43% e.e. (*R*)
R = Bu^t 43% e.e. (*R*)
R = CO_2H 83% e.e. (*R*)

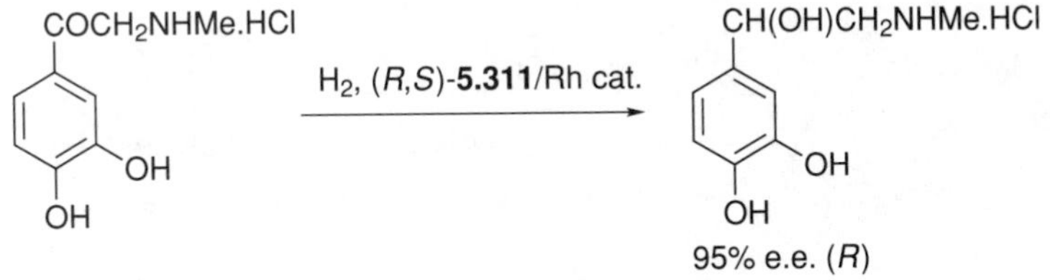

$$3,4\text{-(HO)}_2\text{C}_6\text{H}_3\text{COCH}_2\text{NHMe.HCl} \xrightarrow{\text{H}_2,\ (R,S)\text{-}\mathbf{5.311}\text{/Rh cat.}} 3,4\text{-(HO)}_2\text{C}_6\text{H}_3\text{CH(OH)CH}_2\text{NHMe.HCl} \tag{5.139}$$

95% e.e. (*R*)

$$\text{Ph(HO}_2\text{C)C=CMe}_2 \xrightarrow[\text{100\% yield}]{\text{H}_2,\ 50\ \text{atm.},\ (R,S)\text{-}\mathbf{5.312},\ \text{Rh(NBD)}_2\text{BF}_4\ \text{cat.}} \text{PhCH(CHMe}_2\text{)CO}_2\text{H} \tag{5.140}$$

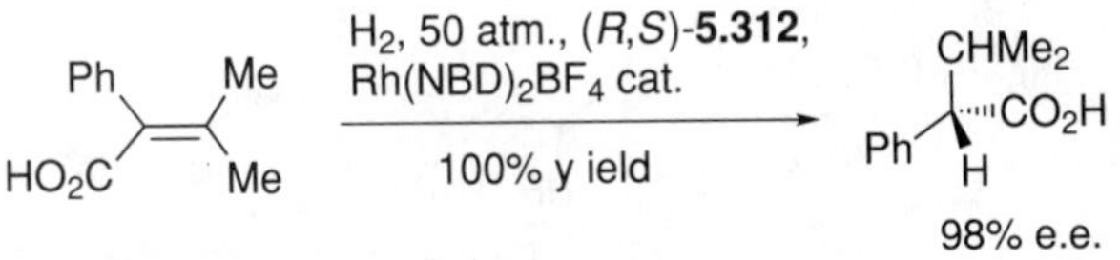

98% e.e.

A number of other applications of these ferrocenylphosphine ligands in palladium-catalyzed reactions have been developed by Hayashi *et al.* Often, the best results are obtained by some modification of the aminoethyl substituent, which illustrates the considerable flexibility of these ligands, since such modifications are relatively easy. This flexibility offsets to some extent the fact that the optically pure complexes must be prepared by resolution, a process that is often considered inelegant in contemporary organic synthesis. In the present cases, however, the resolutions are efficient, they provide both enantiomeric series, and the compounds are not consumed during their application, as would be the case if they were used as stoichiometric intermediates. Other applications are illustrated schematically here, and for the most part are self-explanatory. They include kinetic resolution of allylic acetates (equation 5.141)[104], asymmetric allylic amination and alkytion (equation 5.142 and 5.143)[105–107] asymmetric carbamate cyclizations (equation 5.144)[108], and asymmetric [3+2] cycloaddition reactions (equation 5.145)[109]. The latter is based on Trost's[110] method using trimethylenemethane–palladium complexes, generated catalytically. In most of these conversions, comparison with other commonly used chiral phosphine ligands reveals that the ferrocenyl derivatives are superior.

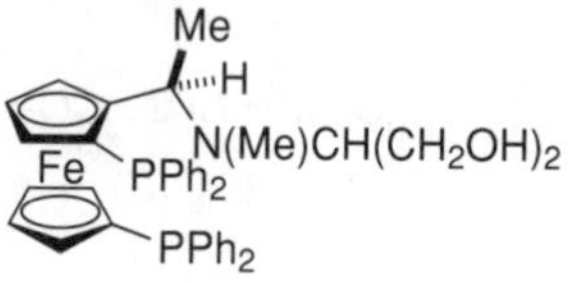

Me, H, N(Me)CH(CH$_2$OH)$_2$, Fe, PPh$_2$, PPh$_2$ **5.314**

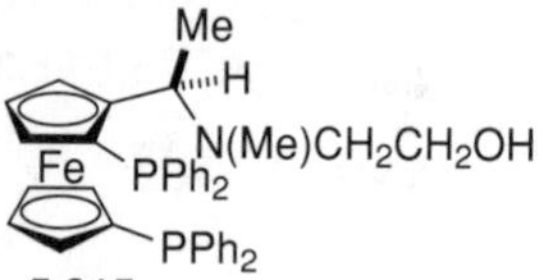

Me, H, N(Me)CH$_2$CH$_2$OH, Fe, PPh$_2$, PPh$_2$ **5.315**

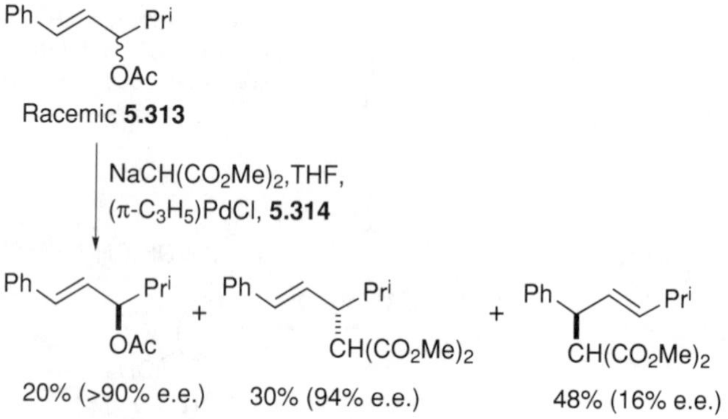

Ph, Pr^i, OAc
Racemic **5.313**

NaCH(CO$_2$Me)$_2$,THF, (π-C$_3$H$_5$)PdCl, **5.314**

Ph, Pr^i, OAc + Ph, Pr^i, CH(CO$_2$Me)$_2$ + Ph, Pr^i, CH(CO$_2$Me)$_2$ (5.141)

20% (>90% e.e.) 30% (94% e.e.) 48% (16% e.e.)

Me, OAc → PhCH$_2$NH$_2$, Pd/**5.314**, 0°C, 90h; 87% yield → Me, NHCH$_2$Ph + Me, NHCH$_2$Ph (5.142)

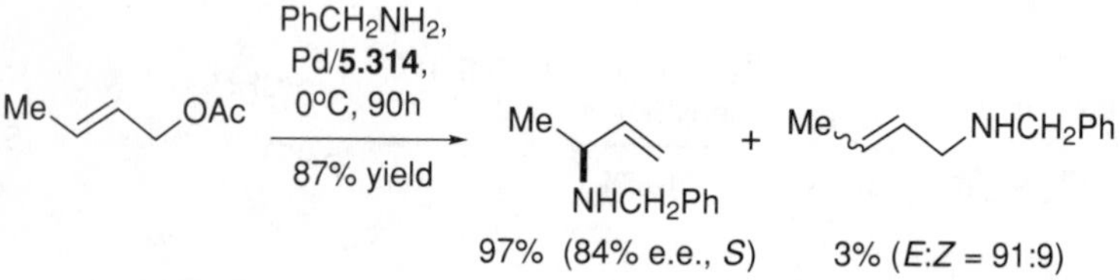

97% (84% e.e., *S*) 3% (*E*:*Z* = 91:9)

1) NaH, THF
2) $CH_2=CHCH_2OAc$, $(\pi\text{-}C_3H_5)PdCl/$**5.315**
100% yield
73% e.e. (*S*)
(5.143)

PhNHCOO–CH₂CH=CHCH₂–OCONHPh → Pd/**5.314**, THF, Δ, 2h; 80% yield
73% e.e. (*S*)
(5.144)

Pd/**5.314**, THF, Δ, 40h; 58% yield
82% (73% e.e.) 18% (58% e.e.)
(5.145)

The development of the ferrocenylphosphine ligands discussed above is perhaps unanticipated, since the connection with ferrocene, and its conversion into substituted derivatives using a variety of techniques, is not obvious. The recognition of these systems as potential ligands for asymmetric transition metal-catalyzed reactions represents one of the creative steps that is necessary before a host of applications in synthesis emerge. We can surely anticipate further developments of this nature in many areas of organometallic chemistry.

5.3 EXPERIMENTAL PROCEDURES

Some of the preparations described in this chapter are multistep operations.

Preparation of triphenylmethyl (trityl) tetrafluoroborate or hexafluorophosphate

Trityl cation is used as a hydride abstraction reagent in organometallic chemistry, especially for the conversion of diene into dienyliron complexes. The procedure given here is for *ca* 10 g of the tetrafluoroborate salt, but the hexafluorophosphate is prepared in an identical fashion by substituting commercially available aqueous hexafluorophosphoric acid (*ca* 65%) for the tetrafluoroboric acid. For many complexes the hexafluorophosphates are preferred owing to their ease of crystallization, but this must usually be established by experiment. Trityl salts are commercially available but the author prefers to make them and use them fresh, since they are very reactive toward moisture and usually do not have a good shelf life. If the following procedure does not give good yield of high quality product, it is usually because the triphenylmethanol is impure. Recrystallization of this compound from methanol affords pure material.

Triphenylmethanol (8.0 g) is dissolved in acetic anhydride (55 ml; propionic anhydride often gives better results) in a round bottom flask by heating on a steam bath. The solution is magnetically stirred and cooled in an ice–acetone (−10°C) bath. For large-scale preparations an overhead stirrer is recommended. Aqueous tetrafluoroboric acid (*ca* 48%, 9.2 ml) is added dropwise, while maintaining the temperature of the reaction mixture below 20°C (lower yields of poor quality material are obtained if this is not observed). After addition of the acid, stirring is continued for 15–20 min and the mixture is poured into dry diethyl ether (*ca* 250 ml) to precipitate the trityl tetrafluoroborate. The product is removed by filtration at the pump and is washed thoroughly with *dry* diethyl ether. The product is very moisture sensitive and the filtration and washing must be carried out in the shortest possible time. The yellow residue is dried under vacuum to remove traces of solvent and may be stored in a screw cap reagent jar for short periods, preferably in a desiccator.

*Preparation of tricarbonyl(4-methoxy-1-methylcyclohexadienyl)iron hexafluorophosphate (**5.9**)*

This preparation represents a typical procedure for hydride abstraction from cyclic diene-$Fe(CO)_3$ complexes using trityl hexafluorophospate[111]. Tricarbonyl(1-methoxy-4-methylcyclohexadiene)iron (**5.8**, 140 g, prepared as in Chapter 4, p. 89) is stirred in methylene chloride (500 ml) at room temperature while trityl hexafluorophosphate (225 g) in methylene chloride (1 l) is added. The mixture is stirred at room temperature for 1 h, concentrated to *ca* 500ml by rotary evaporation, then added to "wet" diethyl ether (*ca* 1.5 l). The product is removed by filtration at the pump and is washed thoroughly with "wet" ether followed by drying under vacuum to give **5.9** (213 g, 95%). The use of "wet" ether (ether previously shaken with a small amount of water and separated) during the isolation of the complex removes excess trityl hexafluorophosphate by hydrolysis to triphenylmethanol. IR (CH_3CN) 2105, 2050 cm^{-1}. The carbonyl frequencies are usually 100 cm^{-1} higher than diene-$Fe(CO)_3$ complexes. ^1H NMR (CD_3CN) δ 6.80 (1H, dd, *J* 6.0, 3.5 Hz, 3-H), 5.55 (1H, d, *J* 6.0 Hz, H-2), 3.79 (3H, s, MeO), 3.85 (1H, m, H-5), 3.00 (1H, *endo*-H-6), 2.40 (1H, *exo*-H-6), 1.75 (3H, s, Me).

*Conversion of methoxy-substituted cyclohexadiene complexes into cyclohexadienyl complexes by acid treatment: preparation of tricarbonyl(2-methylcyclohexadienyl)iron hexafluorophosphate (**5.18**)*[7].

This complex may be prepared from pure **5.8** or from the mixture of isomeric complexes **5.8** and **5.23** that is obtained by direct reaction of pentacarbonyliron with the product from Birch reduction of *p*-methylanisole. Concentrated sulfuric acid (1.2 ml) is added dropwise to the mixture of complexes (0.005 mol) cooled in an ice–water bath. For large-scale preparations it is important not to let the temperature of the reaction mixture rise above 10°C during the addition of acid, otherwise mixtures of isomeric dienyl complexes are obtained. After the addition is complete the mixture is set aside at room temperature for 10–15 min, with occasional swirling. Dry ether (50–100 ml) is added and the mixture is swirled, to precipitate the dienyl salt as a gum, which is repeatedly shaken with dry ether (3 × 50 ml) to remove excess acid and any unreacted diene complexes. The residue is dissolved in cold water (*ca* 20 ml) and the solution is washed with ether. Ammonium hexafluorophosphate (approximately 0.01 mol) dissolved in a small amount of water is added and the water-insoluble complex **5.18** is removed by vacuum filtration, washed with cold water and dried under vacuum (yield: 77%). IR (CH_2Cl_2) 2158, 2127, 2058 cm^{-1}. ^1H NMR (CF_3CO_2H) δ 7.22 (1H, d, *J* 5.5 Hz, H-3), 5.87 (1H, t, *J* 6 Hz, H-4), 4.22 (1H, m, H-5), 3.06 (1H, dt, *J* 17, 7 Hz, H-6), 2.29 (3H, s, Me), 2.19 (1H, d, J_{gem} 17 Hz, H-6).

Preparation of dienyl complexes via oxidative cyclization[16]. *Preparation of complex* **5.46**

This procedure is representative of the general oxidative cyclization methodology. Details for the preparation of the starting diene complex are given in ref. [17]. Manganese dioxide (1.0 g) is refluxed in benzene with a Dean–Stark trap overnight. To the cooled mixture is added complex **5.44** (140 mg) in benzene (1 ml), and the mixture is boiled under reflux for 3 h, at which time TLC shows consumption of all starting material. The mixture is cooled, filtered and evaporated to yield complex **5.45** as a yellow oil (110 mg, 79% yield) which can be used in the next step without further purification. The cyclized complex is dissolved in a mixture of methylene chloride (1.5 ml) and acetic anhydride (0.7 ml) and stirred at 0°C while hexafluorophosphoric acid is added dropwise (0.2 ml of a 65% aqueous solution). After stirring for 30 min, the product is precipitated by pouring the reaction mixture into ether (15 ml) and the dienyl complex is removed by filtration, washed with ether, and dried under vacuum to give **5.46** as a yellow solid (160 mg, 90% yield). IR and NMR spectra are as expected for a dienyl-$Fe(CO)_3$ complex (see above examples).

Stabilized enolate addition to complex **5.9**[48]

Sodium hydride dispersion in mineral oil [72 mg (3 mmol) of NaH] is introduced into a three-neck flask equipped with a rubber septum, argon bubbler and magnetic stirrer, and is washed with dry pentane (5 × 5 ml) using a syringe. Dry, oxygen-free THF (10 ml) is introduced via the septum, and a solution of diethyl malonate (480 mg, 3 mmol) in THF (5 ml) is added dropwise to the stirred suspension at room temperature to give a clear solution (or suspension) of diethyl sodiomalonate.

In a separate flask is placed complex **5.9** (500 mg, 1.22 mmol) and THF (10 ml) and the suspension is stirred under argon at 0°C while the enolate solution is added via syringe, until all of the insoluble complex has disappeared (not all of the malonate solution is required). The reaction mixture is poured into water and the product is extracted with light petroleum in the usual way, followed by purification by chromatography on silica gel with 10% ethyl acetate in hexane as solvent. The product is obtained as a yellow oil (500 mg, 89%). IR 2050, 1970, 1750, 1730 cm^{-1}. NMR spectral details are given in ref. [48].

Decomplexation of diene-Fe(CO)$_3$ complexes from nucleophile addition reactions[48]

By far the most generally useful method for removing the $Fe(CO)_3$ group is that developed by Shvo and Hazum, using trimethylamine-*N*-oxide[20]. This reagent leaves acid-sensitive enol ether groups intact, which can be hydrolyzed to ketone at a later stage. The product from the previous preparation (500 mg) is refluxed in dry benzene with trimethylamine-*N*-oxide (1.0 g) for 30 h. (It should be noted that reaction conditions and time required for complete demetallation vary considerably, and each should be established by small-scale experimentation.) The mixture is cooled to room temperature and is filtered through Celite. The filtrate is evaporated and the residue is purified by chromatography on Florisil, eluted with benzene, to afford diethyl 4-methoxy-1-methylcyclohexa-2,4-dienylmalonate (225 mg, 68% yield) as an oil. IR 1755, 1730, 1685, 1655, 1610 cm^{-1}. Hydrolysis to the cyclohexenone is readily accomplished by using oxalic acid in methanol–water at room temperature for *ca* 45 min, followed by standard extractive work up.

Chiral building blocks: synthesis of optically pure tricarbony(1-methoxycarbonylcyclohexadienyl)iron hexafluorophosphate **(5.11)**[75].

Reaction of methyl cyclohexa-2,5-dienecarboxylate with pentacarbonyliron in refluxing di-n-butyl ether, using the method described on p. 89, gives a mixture of isomeric ester-substituted cyclohexadiene-$Fe(CO)_3$ complexes which is converted into racemic complex **5.58** by heating 22 g with sulfuric acid (40 ml) in methanol (200 ml) at reflux temperature under nitrogen for 24 h, followed by removal of *ca* 100 ml of methanol on the rotary evaporator, addition of water (200 ml) and refluxing for a further 30 h. The product is extracted with ether in the usual way to give racemic **5.58** (19.8 g, 95% yield).

(−)-1-Phenylethylamine (9.5 ml, 75 mmol) is added dropwise to a stirred solution of the racemic acid (19 g, 72 mmol) in chloroform/acetone (3:1, 400 ml). After 30 min the precipitate is filtered off and washed with chloroform, and the washings and filtrate are set aside. Recrystallization of the precipitate twice from chloroform gives the pure diastereomer (7.4 g), $[\alpha]_D$ +68° (*c* 1, acetone). Concentration of the combined washings and filtrate followed by cooling at 0°C overnight gives the other diastereomer $[\alpha]_D$ −95° (6.2 g) which is further recrystallized from chloroform to give pure diastereomer (3.3 g, $[\alpha]_D$ −126°). Repetition of the crystallization procedure on combined washings and filtrates gives a further 0.4 g of pure (+)-isomer and 2.6 g of (−)-isomer. Each diastereomeric ammonium salt is dissolved separately in ethanol containing 2 M aqueous HCl, the solution is partitioned between ether and water and the organic phase is separated, washed with 2 M aqueous HCl, water, dried ($MgSO_4$) and evaporated to give the corresponding acids in quantitative yield. The (−)-salt gives the (−)-acid **5.58A**, $[\alpha]_D$ −136° (*c* 0.1, acetone).

An ether solution of the (−)-acid (2.2 g in 20 ml) is treated with diazomethane (0.5 g) in ether (30 ml) at room temperature for 30 min. Formic acid (0.5 ml) is added to destroy any excess diazomethane and the solution is rotary evaporated to give the methyl ester which can be purified by column chromatography (silica gel, 10% EtOAc in hexane). Yield: 2.3 g (100%), $[\alpha]_D$ −115° (*c* 0.3, $CHCl_3$). The ester (2.2 g) in dry hexane (10 ml) is added dropwise to a solution of triphenylmethyl hexafluorophosphate (4 g) in methylene chloride (35 ml). If a precicpitate is formed during the addition, this should be dissolved by adding the minimum amount of methylene chloride. After 3 h the orange precipitate is filtered off and washed with reagent-grade ether. The solid is redissolved in acetone and re-precipitated by addition of ether, to afford pure dienyl complex (−)-**5.11** (2.4 g, 73% yield; *ca* 16% unreacted starting material may also be recovered), $[\alpha]_D$ −162° (*c* 0.3, acetone).

*Preparation of dicarbonylcycloheptadienyl(triphenyl phosphite)iron hexafluorophosphate (***5.257***)*[90]

Cycloheptadiene (47 g, 0.5 mol) is magnetically stirred in di-n-butyl ether (300 ml) in a 1-liter single neck flask while nitrogen is bubbled through the mixture for 15 min, and then pentacarbonyliron (147 g, 0.75 mol) is added (fume hood!). A reflux condenser fitted with a nitrogen bubbler is attached, and the stirred mixture is heated in an oil bath at 150°C for 44 h. The cooled mixture is worked up as described on p. 90 to give tricarbonylcycloheptadieneiron as a yellow oil (109 g, 93% yield).

This complex (100 g) and triphenyl phosphite (145 g, 1.1 equiv.) in di-n-butyl ether (700 ml) is heated with stirring under reflux using a balloon of argon attached to the top of the condenser. Periodic release of CO from the balloon maintains a *slight* positive pressure of carbon monoxide; an inert gas bubbler may be used instead of the balloon, but the yield is usually a few per cent lower. (*Note: CO is evolved during this reaction: it is essential to follow these instructions properly to avoid rupture of the balloon and consequent discharge of the flask contents due to sudden release of pressure.*) After refluxing for 36 h, the mixture is cooled to room temperature, filtered through Celite, and rotary evaporated to give the crude product, which is purified by column chromatography (silica gel, 10% EtOAc in hexane). Pure dicarbonylcycloheptadiene(triphenyl phosphite)iron is obtained as yellow crystals, m.p. 89–91°C (209 g, 95%).

Triphenylmethyl hexafluorophosphate (120 g) is dissolved in the minimum volume of methylene chloride and dicarbonylcycloheptadiene(triphenyl phosphite)iron (150 g) is added. The mixture is swirled to dissolve the complex and set aside at room temperature for 2 h. The product is precipitated by pouring the reaction solution into excess wet ether, collected by filtration and washed with ether to give complex **5.257** (190.0g, 91% yield). IR (CH_3CN) 2067, 2028 cm^{-1}. Details of NMR spectra of all complexes are given in ref. [90].

Reaction of cycloheptadienyl-Fe(CO)$_2$P(OPh)$_3$ cation with dimethylcopperlithium

To a stirred suspension of cuprous iodide (380 mg) in dry diethyl ether (25 ml) at 0°C under nitrogen atmosphere is added dropwise, via syringe, methyllithium (1.8 M in ether). A yellow precipitate of methylcopper forms, which disappears to form a clear solution as more than one equivalent of methyllithium is added. Sufficient lithium reagent is added to just dissolve the methylcopper. The solution is stirred for 5–10 min at 0°C, and the cycloheptadienyl complex **5.257** (1.3 g) is added via a powder addition funnel. The reaction mixture is stirred for 10–15 min at 0°C and is then added to saturated aqueous ammonium chloride (50 ml) and stirred vigorously in air for 30 min. Insoluble copper salts are removed by filtration, the filter paper is washed with ether and the combined ether extracts are dried ($MgSO_4$) and rotary evaporated to give the product which is purified by chromatography (silica gel, 10% EtOAc in hexane). Yield: 1.03 g, 97%, m.p. 87.5–88.5°C. IR ($CHCl_3$) 1990, 1938 cm^{-1}.

Synthesis of optically pure α-ferrocenylethyldimethylamine (**5.300**)[100]

The starting material for this preparation, α-ferrocenylethanol, is prepared by reduction of acetylferrocene with lithium aluminum hydride[112]. A 3-neck 1-liter flask fitted with a reflux condenser, nitrogen inlet, dropping funnel and magnetic stir bar, is charged with acetylferrocene (22.8 g) in dry ether (500 ml). To the stirred solution is added dropwise a solution of lithium aluminum hydride (1.9 g) in ether, and the mixture is heated under reflux for 2 h. Excess lithium aluminum hydride is destroyed with ethyl acetate and the reaction mixture is treated with aqueous ammonium chloride (26.8 g of NH_4Cl). After stirring for 30 min at 0°C, the mixture is filtered and the organic layer is separated, washed with water (2 × 50 ml), dried ($MgSO_4$) and rotary evaporated. The crude product (20.5 g, 89%) has m.p. 69–72°C and may be used in the next step without further purification. Recrystallization from ether/petroleum ether affords orange rods, m.p. 73–75°C.

A solution of α-ferrocenylethanol (23.0 g) in toluene (150 ml) is added dropwise to a stirred solution of phosgene (12.5 g) in toluene (100 ml) at −20°C. After stirring for 30 min, the solution is allowed to warm to +20°C and is added dropwise to a cooled (−20°C) solution of dimethylamine (22.5 g) in isopropanol (200 ml). The reaction mixture is filtered, evaporated to dryness, and the residue is taken up in benzene. The amine is extracted into 8.5% phosphoric acid, washed with benzene, the acid extract is basified with sodium carbonate and the amine is extracted with benzene, dried ($MgSO_4$) and evaporated to give the crude product (24.4 g, 95%). The product is purified by distillation (b.p. 110°C at 0.45 mm Hg, some decomposition is observed) to give racemic complex **5.300** (17.5 g, 68 % yield), but the crude product may be used in the optical resolution.

Racemic **5.300** (51.4 g) and (*R*)-(+)-tartaric acid (30.0 g) are each dissolved in methanol (100 ml) and the solutions are mixed at 55°C while being stirred. Seeding crystals are added and the temperature is slowly lowered (2° h^{-1}). After 24 h, the mixture is filtered to yield 30.0 g of partially resolved tartrate which is treated with base and extracted with benzene to give optically active **5.300** ($[\alpha]_D$ −11.0°, *c* 1.5, ethanol). This material is dissolved in methanol (50 ml) and combined with (*R*)-(+)-tartaric acid (11.1 g) in methanol (50 ml) at 55°C. Seeding, followed by slow cooling as before gives the tatrate salt (27.5 g) which is converted into (*S*)-(−)-**5.300** (17.0 g, 66% yield. $[\alpha]_D$ −14.1°, c 1.6, ethanol). The (*R*)-enantiomer is obtained from the mother liquors of the first crystallization by concentration to 25% of its original volume and addition of diethyl ether until no further precipitate is formed. The mixture is set aside at 0°C overnight and the precipitate is collected (48.6 g) and recrystallized from acetone–water (10:1, 800 ml) to give 34.5 g of tartrate which is recrystallized again from 500 ml of aqueous acetone (yield: 28.0 g). The tartrate is converted into (*R*)-**5.300** in the usual way, $[\alpha]_D$ +14°. The mother liquors can be recycled to give an overall recovery of 80–90% of both antipodes in optically pure form.

References

1. E. O. Fischer and R. D. Fischer, *Angew. Chem.* **72**, 919 (1960).
2. O. Eisenstein, W. Butler and A. J. Pearson, *Organometallics* **3** 1150 (1984).
3. A. J. Birch, K. B. Chamberlain, M. A. Haas and D. J. Thompson, *J. Chem. Soc. Perkin Trans.* **1**, 1882 (1973).
4. R. E. Ireland, G. G. Brown, Jr, R. H. Stanford, Jr and T. C. McKenzie, *J. Org. Chem.* **39**, 51 (1974).
5. A. J. Birch and D. H. Williamson, *J. Chem. Soc. Perkin Trans.* **1**, 1892 (1973).
6. L. A. Paquette, R. G. Daniels and R. Gleiter, *Organometallics* **3**, 560 (1985).
7. A. J. Birch and M. A. Haas, *J. Chem. Soc. (C)* 2465 (1971).
8. A. J. Birch and W. D. Raverty, unpublished data, Australian National University, 1976.
9. P. W. Howard, G. R. Stephenson and S. C. Taylor, *J. Chem. Soc. Chem. Commun.* 1603 (1988).
10. P. W. Howard, G. R. Stephenson and S. C. Taylor, *J. Chem. Soc. Chem. Commun.* 1182 (1990).
11. G. R. Stephenson, P. W. Howard and S. C. Taylor, *J. Chem. Soc. Chem. Commun.* 127 (1991).
12. P. W. Howard, G. R. Stephenson and S. C. Taylor, *J. Organomet. Chem.* **370**, 97 (1989).
13. D. R. Boyd, M. R. Dorrity, M. V. Hand, J. F. Malone, N. D. Sharma, H. Dalton, D. J. Gray and G. N. Sheldrake, *J. Am. Chem. Soc.* **113**, 666 (1991).
14. A. J. Pearson, A. M. Gelormini and A. A. Pinkerton,*Organometallics* **11**, 936 (1992).
15. A. J. Birch, K. B. Chamberlain and D. J. Thompson, *J. Chem. Soc. Perkin Trans.* **1**, 1900 (1973).
16. A. J. Pearson, *J. Chem. Soc. Chem Commun.* 488 (1980).
17. A. J. Pearson and C. W. Ong, *J. Org. Chem.* **47**, 3780 (1982).
18. E. G. Bryan, A. L. Burrows, B. F. G. Johnson, J. Lewis and G. M. Schiavon, *J. Organomet. Chem.* **127**, C19 (1977).

19. R. J. H. Cowles, B. F. G. Johnson, P. L. Josty and J. Lewis, *J. Chem. Soc. Chem. Commun.* 392 (1962).
20. Y. Shvo and E. Hazum, *J. Chem. Soc. Chem. Commun.* 336 (1974).
21. K. Nunn, P. Mosset, R. Grée and R. W. Saalfrank, *Angew Chem. Int. Ed. Engl* **27**, 1188 (1988).
22. D. J. Thompson, *J. Organomet. Chem.* **108**, 381 (1976).
23. D. H. R. Barton, A. A. L. Gunatilaka, T. Nakanishi, H. Patin, D. A. Widdowson and B. R. Worth, *J. Chem. Soc. Perkin Trans.* **1**, 821 (1976).
24. G. R. Stephenson, *J. Chem. Soc. Perkin Trans.* **1**, 2449 (1982).
25. A. J. Birch and G. R. Stephenson, *J. Organomet. Chem.* **218**, 91 (1981).
26. A. J. Birch, W. D. Raverty and G. R. Stephenson, *J. Org. Chem.* **46**, 5166 (1981).
27. A. J. Birch and B. M. R. Bandara, *Tetrahedron Lett.* **21**, 2981 (1980).
28. J. A. S. Howell and M. J. Thomas, *J. Chem. Soc. Dalton Trans.* 1401 (1983).
29. A. J. Birch, W. D. Raverty and G. R. Stephenson, *Organometallics* **3**, 1075 (1984).
30. A. J. Pearson, in *Comprehensive Organometallic Chemistry* (ed. G. Wilkinson, F. G. A. Stone and E. W. Abel), Vol. 8, Chapter 58, Pergamon Press, Oxford, 1982.
31. A. J. Pearson, *Second Supplements to the 2nd Edition of Rodd's Chemistry of Carbon Compounds, Vol. II A and B (Partial)* (ed. M. Sainsburg), Chapter 5c, Elsevier, Amsterdam, 1992.
32. D. Astruc, in *The Chemistry of the Metal–Carbon Bond, Vol. 4*, (ed. F. A. Hartley), Chapter 7, Wiley, Chichester, 1987.
33. A. J. Pearson in *The Chemistry of Metal–Carbon Bond,* Vol. 4, (ed. F. A. Hartley), Chapter 10, Wiley, Chichester, 1987.
34. A. J. Pearson, *Advances in Metal-Organic Chemistry*, Vol. 1, Chapter 1, JAI Press, Inc., Greenwich, 1989.
35. A. J. Birch, P. E. Cross, J. Lewis, D. A. White and S. B. Wild, *J. Chem. Soc., (A)*, 332 (1968).
36. A. J. Birch and A. J. Pearson, *J. Chem. Soc. Perkin Trans.* **1**, 954 (1976).
37. A. J. Pearson, *Aust. J.Chem.* **29**, 1101 (1976).
38. B. M. R. Bandara, A. J. Birch and T-C. Khor, *Tetrahedron Lett.* **21**, 3625 (1980).
39. A. J. Pearson and J. Yoon, *Tetrahedron Lett.* **26**, 2399 (1985).
40. A. J. Birch, A. J. Liepa and G. R. Stephenson, *Tetrahedron Lett.* 3565 (1979).
41. G. R. John and L. A. P. Kane-Maguire, *J. Chem. Soc. Dalton Trans.* 1196 (1979).
42. G. A. Potter and R. McCague, *J. Chem. Soc. Chem. Commun.* 635 (1992).
43. D. Seebach, *Angew. Chem. Int. Ed. Engl.* **18**, 239 (1979).
44. A. J. Pearson and D. C. Rees, *J. Am. Chem. Soc.* **104**, 1118 (1982).
45. A. J. Pearson and D. C. Rees, *J. Chem. Soc. Perkin Trans.* **1**, 2467 (1982).
46. A. J. Pearson, D. C. Rees and C. W. Thornber, *J. Chem. Soc. Perkin Trans.* **1**, 619 (1983).
47. G. Stork and J. E. Dolfini, *J. Am. Chem. Soc.* **85**, 2872 (1963).
48. A. J. Pearson, *J. Chem. Soc. Perkin Trans.* **1**, 2069 (1977).
49. A. J. Pearson, T. R. Perrior and D. C. Rees, *J. Organomet. Chem.* **226**, C 39 (1982).
50. B. M. Trost and T. R. Verhoeven, *J. Am. Chem. Soc.* **102**, 4743 (1980).
51. P. A. Krapcho, J. F. Weimaster, J. M. Eldridge, E. G. E. Jahngen, A. J. Lovey and W. P. Stevens, *J. Org. Chem.* **43**, 138 (1978).
52. A. J. Pearson, P. Ham, C. W. Ong, T. R. Perrior and D. C. Rees, *J. Chem. Soc. Perkin Trans.* **1**, 1527 (1982).
53. A. J. Pearson and M. K. O'Brien, *J. Org. Chem.* **54**, 4663 (1989) and references cited therein.
54. Ch. Tamm, *Fortschr. Chem. Org. Naturst.* **31**, 63 (1974).
55. P. G. McDougal and R. N. Schmuff, *Prog. Chem. Org. Nat. Prod.* **47**, 153 (1985).
56. C. J. Miroca, S. V. Pathre and C. M. Christenson, *Mycotoxic Fungi, Mycotoxins, Mycotoxicoses.*, Vol. 1, pp. 365–409, Dekker, New York, 1977.
57. A. J. Birch, L. F. Kelly and A. S. Narula, *Tetrahedron* **38**, 1813 (1982).
58. I. Fleming and P. E. Sanderson, *Tetrahedron Lett.* **28**, 4229 (1987).
59. K. Tamao, N. Ishida, T. Tanaka and M. Kumada, *Organometallics* **2**, 1694 (1983).
60. For other syntheses of (±)-trichodermol, see: E. W. Colvin, S. Malchenko, R. A. Raphael and J. S. Roberts, *J. Chem. Soc. Perkin Trans.* **1**, 1989 (1973), for other syntheses of (±)-trichodermol.

61. W. C. Still and M. T. Tsai, *J. Am. Chem. Soc.* **102**, 3654 (1980), for other syntheses of (±)-trichodermol.
62. A. J. Pearson, *J. Chem. Soc. Perkin Trans.* **1**, 1255 (1979).
63. J. A. Marshall, S. F. Brady and N. H. Andersen, *Fortschr. Chem. Org. Naturst.* **31**, 283 (1974).
64. A. J. Pearson, P. Ham and D. C. Rees, *Tetrahedron Lett.* **21**, 4637 (1980).
65. A. J. Pearson, P. Ham and D. C. Rees, *J. Chem. Soc. Perkin Trans.* **1**, 489 (1982).
66. J. W. Daly, I. Karle, C. W. Myers, T. Tokuyama, J. A. Waters and B. Witkop, *Proc. Natl. Acad. Sci. USA.* **68**, 1870 (1971).
67. A. J. Pearson and P. Ham, *J. Chem. Soc. Perkin Trans.* **1**, 1421 (1983), and references cited therein.
68. A. J. Pearson and T. R. Perrior, *J. Organomet. Chem.* **285**, 253 (1985).
69. A. J. Pearson, *J. Chem. Soc. Perkin Trans.* **1**, 400 (1980).
70. H. J. Knölker, *Synlett* 371 (1992).
71. A. J. Pearson, I. C. Richards and D. V. Gardner, *J. Org. Chem.* **49**, 3887 (1984).
72. G. R. Stephenson, D. A. Owen, H. Finch and S. Swanson, *Tetrahedron Lett.* **32**, 1291 (1991).
73. I. H. Sanchez and F. R. Tallabs, *Chem Lett.* 891 (1981).
74. A. J. Birch, B. M. R. Bandara and K. Chamberlain, *Tetrahedron* **37**, W289 (1981). (Woodward Memorial Issue).
75. B. M. R. Bandara, A. J. Birch and L. F. Kelly, *J. Org. Chem.* **49**, 2496 (1984).
76. A. J. Birch, L. F. Kelly and D. V. Weerasuria, *J. Org. Chem.* **53**, 278 (1988).
77. P. A. Grieco and S. D. Larsen, *J. Org. Chem.* **51**, 3553 (1986).
78. A. J. Pearson, T. Ray, I. C. Richards, J. C. Clardy and L. Silveira, *Tetrahedron Lett.* **24**, 5827 (1983).
79. W. A. Donaldson and M. Ramaswamy, *Tetrahedron Lett.* **30**, 1343 (1989).
80. W. A. Donaldson and M Ramaswamy, *Tetrahedron Lett.* **29**, 1343 (1988).
81. R. S. Bayoud, E. R. Biehl and P. C. Reeves, *J. Organomet. Chem.* **174**, 297 (1979).
82. R. S. Bayoud, E. R. Biehl and P. C. Reeves, *J. Organomet. Chem.* **150**, 75 (1978).
83. G. Maglio and R. Palumbo, *J. Organomet. Chem.* **76**, 367 (1974).
84. B. Samuelsson, *Science* **220**, 568 (1983).
85. P. Borgeat and P. Sirois, *J. Med. Chem.* **24**, 121 (1981).
86. W. A. Donaldson and M. Ramaswamy, *Tetrahedron Lett.* **30**, 1339 (1989).
87. R. Aumann, *J. Organomet. Chem.* **47**, C28 (1973).
88. R. Edwards, J. A. S. Howell, B. F. G. Johnson and J. Lewis, *J. Chem. Soc. Dalton Trans.* 2105 (1974).
89. B. F. G. Johnson, J. Lewis, T. W. Matheson, I. E. Ryder and M. V. Twigg, *J. Chem. Soc. Chem. Commun.* 269 (1974).
90. A. J. Pearson, S. L. Kole and T. Ray, *J. Am. Chem. Soc.* **106**, 6060 (1984).
91. A. J. Pearson, Y. S. Lai, W. Lu and A. A. Pinkerton, *J. Org. Chem.* **54**, 3882 (1989).
92. J. E. Bäckvall, S. E. Byström and R. E. Nordberg, *J. Org. Chem.* **49**, 4619 (1984).
93. T. J. Kealy and P. L. Pauson, *Nature (London)* **168**, 1039 (1951).
94. B. Rockett and G. Marr, *J. Organomet. Chem.* Annual Reports.
95. *Gmelin*, Handbüch der Anorganischen Chemie. New Supplementary Series A, Vol. 14; Vol. 41; Vol. 49; Vol. 50.
96. F. C. Hedberg and H. Rosenberg, *Tetrahedron Lett.* 4011 (1969).
97. M. D. Rausch, *Inorg. Chem.* **1**, 227 (1962).
98. M. D. Rausch and D. J. Ciappenelli, *J. Organomet. Chem.* **10**, 127 (1967).
99. F. Rebiera, O. Samuel and H. B. Kagan, *Tetrahedron Lett.* **31**, 3121 (1990).
100. D. Marquarding, H. Klusacek, G. Gokel, P. Hoffmann and I. Ugi, *J. Am. Chem. Soc.* **92**, 5389 (1970).
101. C. R. Hauser and J. K. Lindsay, *J. Org. Chem.* **22**, 906 (1957).
102. T. Hayashi and M. Kumada, *Acc. Chem. Res.* **15**, 395 (1982).
103. V. I. Sokolov, L. L. Troitskaya and O. A. Reutov, *J. Organomet. Chem.* **202**, C58 (1980).
104. T. Hayashi, A. Yamamoto and Y. Ito, *J. Chem. Soc. Chem. Commun.* 1090 (1986).

105. T. Hayashi, K. Kishi, A. Yamamoto and Y. Ito, *Tetrahedron Lett.* **31**, 1743 (1990).
106. T. Hayashi, A. Yamamoto, Y. Ito, E. Noshioka, H. Miura and K. Yanagi, *J. Am. Chem. Soc.* **111**, 6301 (1989).
107. T. Hayashi, K. Kanehira, T. Hagihara and M. Kumada, *J. Org. Chem.* **53**, 113 (1988).
108. T. Hayashi, A. Yamamoto and Y. Ito, *Tetrahedron Lett.* **29**, 99 (1988).
109. A. Yamamoto, Y. Ito and T. Hayashi, *Tetrahedron Lett.* **30**, 375 (1989).
110. B. M. Trost, *Angew. Chem. Int. Ed. Engl.* **25**, 1 (1986).
111. A. J. Pearson and C. W. Ong, *J. Am. Chem. Soc.* **103**, 6686 (1981).
112. F. S. Arimoto and A. C. Haven, Jr, *J. Am. Chem. Soc.* **77**, 6295 (1955).

–6–

Arene–Iron Complexes

6.1 INTRODUCTION

Because of the general importance of aromatic compounds in organic chemistry, both as intermediates in the synthesis of more complex molecules and as targets themselves with desirable properties, the chemistry of arene–metal complexes has received much attention as a means to functionalize aromatic molecules. Perhaps the most familiar of these types of complex are the arene chromium tricarbonyl derivatives[1]. A transition metal attached to the aromatic ring confers a number of useful properties on the ligand, leading to methodology for the synthesis of variously substituted aromatic molecules from simple precursors. Arene–iron complexes have been investigated in some detail, but have lagged behind the chromium compounds in terms of the variety of carbon–carbon bond forming reactions that have been exploited for synthetic purposes. This is unfortunate, since the iron complexes, being positively charged, are more reactive toward nucleophiles and have much to offer in terms of flexibility. There are a number of problems associated with the manipulation of, for example, products of nucleophile addition reactions, and there is still much to be done before applications of this chemistry to sophisticated organic synthesis become commonplace.

There are two types of arene–iron complex that can be prepared and studied: arene-FeCp and bis(arene)Fe complexes. Both of these may be thought of as complexes of Fe(II), although arene-FeCp complexes are often referred to as Fe(I) complexes, since the cyclopentadienyl ligand is considered a five electron donor. This classification is somewhat arbitrary, and the organic chemist may be more comfortable with the cyclopentadienyl anion, an aromatic 6-electron species. By this reckoning the arene-FeCp complexes require Fe(II). The differences in chemistry observed for arene-FeCp vs. $(arene)_2Fe$ complexes may be rationalized on the basis of overall positive charge, the cyclopentadienyl derivatives being monocations, whereas the bis(arene) systems are dications. Thus, the latter are more reactive toward nucleophiles and can react sequentially with two nucleophiles before the neutral complex stage is reached. On the other hand, arene-FeCp complexes and their nucleophile addition products are more stable, easier to prepare and easier to handle. Indeed, the arene-FeCp derivatives are quite robust, even permitting oxidation of attached methyl substituents to carboxylic acids, as will be shown later.

The general reaction chemistry of arene–metal complexes can be summarized diagrammatically (Fig. 6.1). Owing to their positively charged nature, not all of these characteristics apply to the iron complexes to be discussed in this chapter. For example, whereas arene-$Cr(CO)_3$ systems give stabilization of benzylic carbocations *and* benzylic carbanions, the arene–iron complexes will only support benzylic carbanions. Furthermore, owing to their very high reactivity toward carbanionic species, lithiations of the aromatic nucleus is not a useful reaction with the organoiron systems.

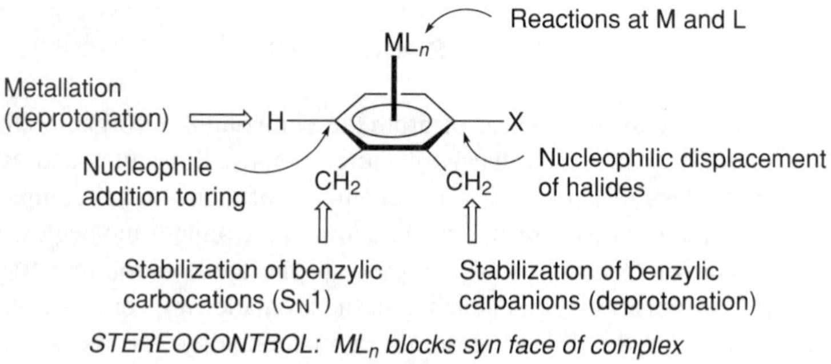

Reactions at M and L
ML_n
Metallation (deprotonation)
H
X
Nucleophile addition to ring
CH_2
CH_2
Nucleophilic displacement of halides
Stabilization of benzylic carbocations (S_N1)
Stabilization of benzylic carbanions (deprotonation)
STEREOCONTROL: ML_n blocks syn face of complex

FIG. 6.1. General reactivity patterns of arene–metal π-complexes.

Since our interest here is in the applications of these complexes in organic synthesis, and the metal and ancillary ligands (ML_n) can be regarded as spectators of events occurring at the arene ligand, we shall not be concerned with reactions that occur directly at M or L in Fig. 6.1, except in those cases where this leads to an eventual alteration of the arene.

This chapter is divided into two sections, one dealing with arene-FeCp complexes and the other with $(arene)_2Fe$ derivatives. Owing to the larger amount of work that has been done on the former compounds, they will occupy most of the chapter. An extensive review by Astruc[2] was published in 1983, to which the reader is referred for a thorough literature survey.

6.2 ARENE-FeCp COMPLEXES

6.2.1 Preparation

These compounds were first prepared via the treatment of an aromatic compound with $CpFe(CO)_2Cl$ in the presence of aluminum chloride (equation 6.1)[3]. This route can also be used for the pentamethylcyclopentadienyl complexes ($Cp^* = C_5Me_5$). Usually the haloaluminate complexes thus obtained are converted into hexafluorophosphates by treatment with aqueous NH_4PF_6, as these derivatives are easier to store and handle.

$$\text{R-C}_6\text{H}_5 \xrightarrow[\text{AlCl}_3,\ \text{heat}]{\text{C}_5\text{H}_5\text{Fe(CO)}_2\text{Cl or C}_5\text{Me}_5\text{Fe(CO)}_2\text{Br,}} [(\text{R-C}_6\text{H}_5)\text{FeCp (Cp*)}]^+ \text{X}^- \quad (6.1)$$

Ferrocene has also been used as a source of FeCp in this procedure, but the ferricinium cation is generated under the conditions shown in equation 6.1, leading to separation and purification problems. Incorporation of aluminum powder into the reaction mixture effects reduction of ferricinium back to ferrocene, and in many cases good yields of arene complex are obtained by this method. Owing to the ready availability and low cost of ferrocene, this is now the most commonly used method for the preparation of the complexes (equation 6.2). For inexpensive arenes, the substrate can be used as solvent, or if the arene is less readily available or a solid, the reaction may be carried out in an inert solvent, usually decalin, cyclohexane, methylcyclohexane, octane or heptane. Sometimes yields can be increased (surprisingly) by introducing one equivalent of water into the reaction brew.

$$\text{R-C}_6\text{H}_5 \xrightarrow[\text{2) NH}_4\text{PF}_6\text{, H}_2\text{O}]{\text{1) Ferrocene, 2AlCl}_3\text{, Al, 70– 190°C, 1–16 h}} [(\text{R-C}_6\text{H}_5)\text{FeCp}]^+ \text{PF}_6^- \quad (6.2)$$

The yields of complex are affected by the nature of the substituent(s) (R in equation 6.2). Electron releasing substituents increase the yield, whereas electron-withdrawing groups lead to yield decrease, often severe. For example, a 90% yield of benzene-FeCp can be obtained, the yield drops to 45% for fluorobenzene and to between 6 and 9% for the isomeric difluorobenzenes. Trifluorobenzenes are completely resistant to complexation. Complexation of halogenobenzenes is somewhat more problematic in terms of side reactions than alkylbenzenes, owing to the tendency for dehalogenation to occur under the usual reaction conditions. In some cases this can be avoided by simple omission of the Al powder and/or use of lower reaction temperatures. In many cases, the analogous arene-RuCp complexes provide a useful alternative, often giving better yields with highly electron-deficient arenes. Also, the method of preparation {[CpRu($CH_3CN)_3]PF_6$, dichloroethane, reflux 2–6h]} is much kinder to sensitive functional groups. Unfortunately, the cost of ruthenium is prohibitive for large-scale stoichiometric applications, and for this reason alone the development of ruthenium-catalyzed transformations is a highly desirable though as yet unrealized goal.

Problems arise during the complexation of condensed polyaromatics owing to the accompanying reduction of one or more of the aromatic rings, although this can sometimes be avoided by using lower temperatures (equation 6.3). Complexation of tetralins and indanes, on the other hand, proceeds satisfactorily.

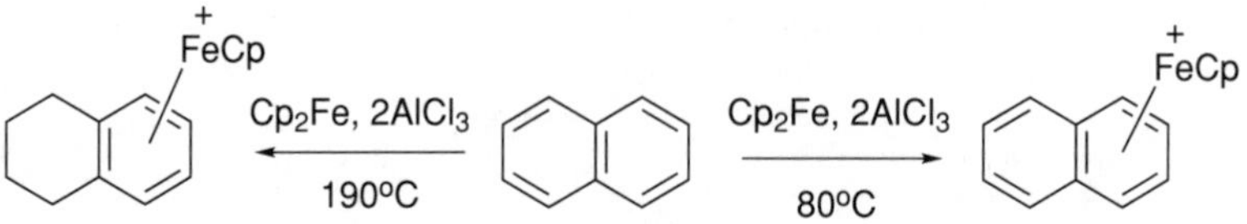

(6.3)

A large number of arene-FeCp complexes have been prepared but will not be discussed here, since we wish to present the salient features of their chemistry that might be exploited in organic synthesis. For an overview of the types of complex available, the reader should consult ref. [2].

6.2.2 Nucleophile addition reactions

Reactions of arene-FeCp complexes occur with stereo-, regio-, and chemoselectivity. In accordance with the Davies, Green, Mingos rules[4], attack occurs at the arene ring except in a few special cases where severe steric hindrance forces the nucleophile to attack at the cyclopentadienyl ring (equations 6.4 and 6.5).

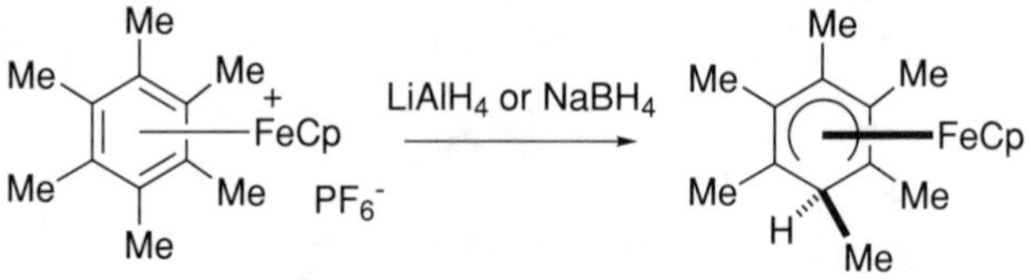

(6.4)

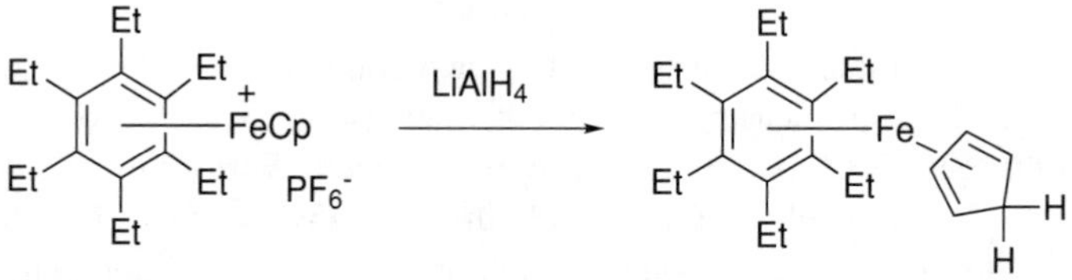

(6.5)

When the nucleophile adds to the arene ring it does so stereoselectively *anti* to the metal, as with other electrophilic 18 electron (olefin)iron complexes, although the stereochemistry is inconsequential at this point since it is never translated into the organic product during decomplexation (see later). With monosubstituted benzene derivatives nucleophile addition occurs at an unsubstituted carbon and with regioselectivity that depends on the substituent (electronic or steric effects) and on the nucleophile (steric effects). Results of borohydride reduction illustrating the directing effects of various substituents are shown in equation 6.6 and Table 6.1. Electron-donating substituents direct mostly *meta*, although this is not as pronounced as with arene-$Cr(CO)_3$ complexes, whereas electron-withdrawing groups direct predominately *ortho*. These effects have been rationalized on the basis of charge control[4]. Very little effort has gone into improving the regiocontrol by electron-donating substituents, and a useful approach might be to use sterically more demanding groups such as isopropoxy, which gives beneficial effects with cyclohexadienyl-$Fe(CO)_3$ complexes (see Chapter 5), or to study stronger electron donors such as dialkylamino, which have shown useful regiocontrol during nucleophile additions to arene-$Mn(CO)_3$ cations[5].

R-C6H5-FeCp+ PF6− —NaBH4→ 6.1 + 6.2 + 6.3 (6.6)

Table 6.1

Borohydride reductions of monosubstituted arene-FeCp complexes (equation 6.6)

Substituent (R)	Ratio 1	:	2	:	3
CH_3	1	:	1	:	0.5
Cl	4	:	1	:	0
CH_3O	0.2	:	1	:	0.6
CO_2CH_3	12.7	:	1	:	1

With two substituents on the ring there will be competing effects, but this may not be too great a problem. Again, very little systematic work has been done in this area. Borohydride reduction of the *p*-methylanisole complex **6.4** gives a 4:1 mixture in favor of attack *meta* to the methoxy substituent (equation 6.7).

6.4 —NaBH4→ **6.5** (80%) + **6.6** (20%) (6.7)

Of greater synthetic utility is the addition of carbon nucleophiles (equations 6.8–6.11). A number of alkyllithiums, enolates, etc., have been shown to add to the arene ligand. With the sterically hindered hexamethylbenzene complex, addition occurs instead to the cyclopentadienyl ligand (equation 6.10). In some cases, the cyclohexadienyl complexes resulting from nucleophile addition can be selectively oxidized to a substituted arene complex using trityl cation or *N*-bromosuccinimide (NBS), although this can be problematic, leading to scission of the C–C bond between dienyl and substituent (equation 6.11). Again, electron-withdrawing substituents direct predominantly *ortho*, whereas electron donors (weakly) direct *meta*.

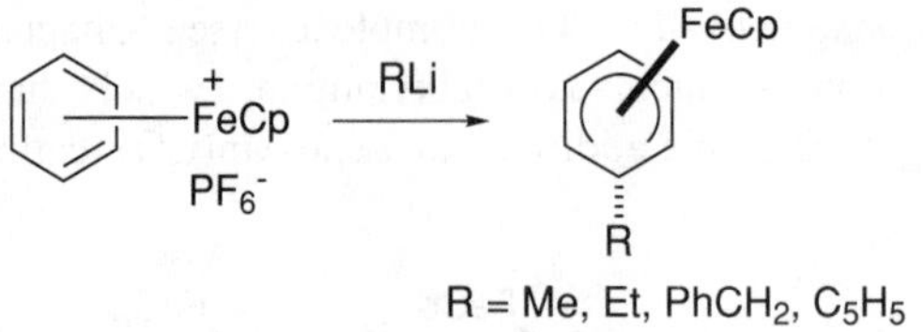

(6.8)

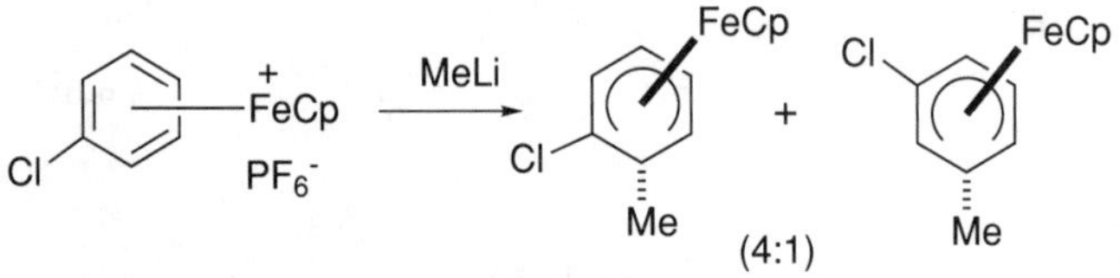

R = Me, Et, $PhCH_2$, C_5H_5

(6.9)

(4:1)

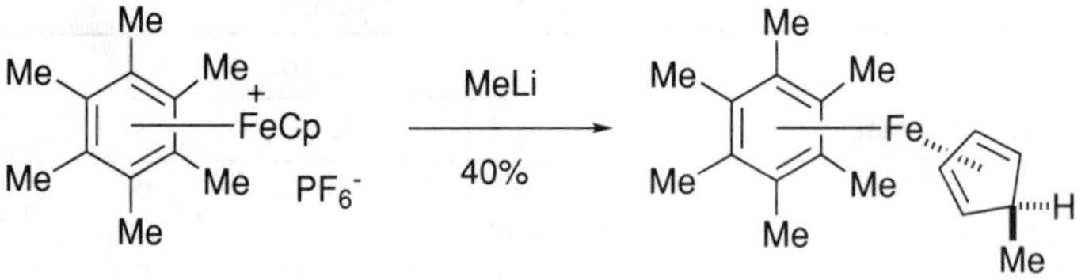

(6.10)

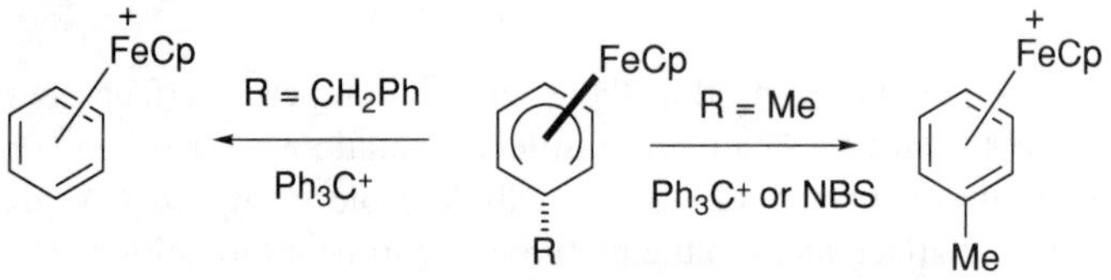

(6.11)

Addition of functionalized carbon nucleophiles is of potentially greater interest in synthesis. In contrast to arene-$Cr(CO)_3$ complexes, relatively unreactive nucleophiles, such as cyanide, nitromethanate and enolates add to the FeCp systems[6,7]. More recently, Sutherland *et al.*[6,7] have studied additions of the trichloromethyl anion, generated from chloroform and potassium t-butoxide[6]. This is potentially useful because the CCl_3 group is readily converted into carboxylic acid and other functional groups. Examples of functional carbon nucleophile additions are given in equations 6.12–6.22. During cyanide reaction with chloroarene complexes, products arising from nucleophile addition *ortho* to the halogen, nucleophilic substitutions of halide (see later), or both pathways, can be obtained, depending on reaction conditions[8]. Short reaction time (3 min) leads almost exclusively to the product of CN addition *ortho* to the chloride (equation 6.15), while a 3 h reaction, followed by filtration of the reaction mixture through a glass sinter, solvent removal, etc., gives the product of substitution and addition (equation 6.16). If the product from extended reaction time is treated with aqueous ammonium hexafluorophosphate during work-up, then the substitution product **6.8** is obtained (equation 6.17). It is known that compounds such as **6.7** lose cyanide on exposure to acid to give arene complexes such as **6.8**, and the effect of work-up conditions is attributed to the acidity of the NH_4PF_6 that is used[8].

FeCp+ ; LiCH(Me)CN ; FeCp ; CH(Me)CN (6.12)

FeCp+ ; X ; CN^-, DMF ; FeCp ; X ; CN

X = NO_2 80%
X = COPh 82%
X = CO_2CH_3 75%
X = SO_2Ph 75% (6.13)

FeCp+ ; X ; O ; CN^-, DMF ; FeCp ; X ; O ; CN

X = CO 85%
X = O 60%
X = S 70%
X = SO_2 85% (6.14)

FeCp+ ; Cl ; CN^-, DMF ; 3 min ; FeCp ; Cl ; CN ; major product (6.15)

FeCp+ ; Cl ; NaCN, DMF ; 3 h; filter through glass sinter ; FeCp ; CN ; CN ; **6.7** (54%) (6.16)

FeCp+ ; Cl ; 1) NaCN, DMF, 3h ; 2) Aq. NH_4PF_6; extract with CH_2Cl_2/CH_3NO_2 ; FeCp+ ; CN ; **6.8** (65%) (6.17)

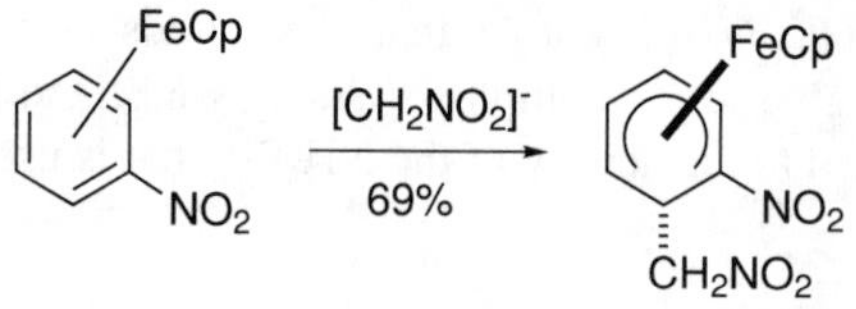

FeCp, NO_2 —[$[CH_2NO_2]^-$, 69%]→ FeCp, NO_2, CH_2NO_2 (6.18)

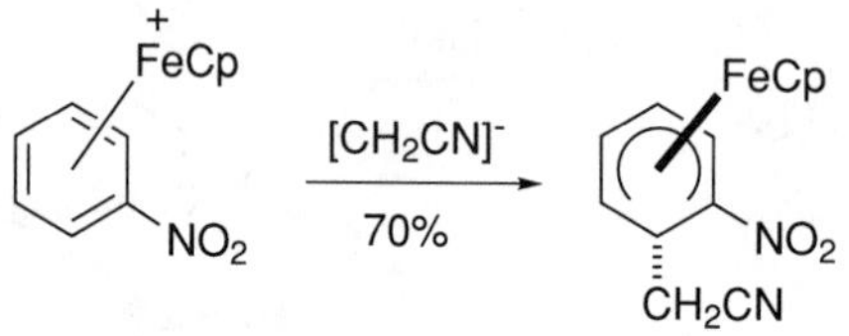

$\overset{+}{FeCp}$, NO_2 —[$[CH_2CN]^-$, 70%]→ FeCp, NO_2, CH_2CN (6.19)

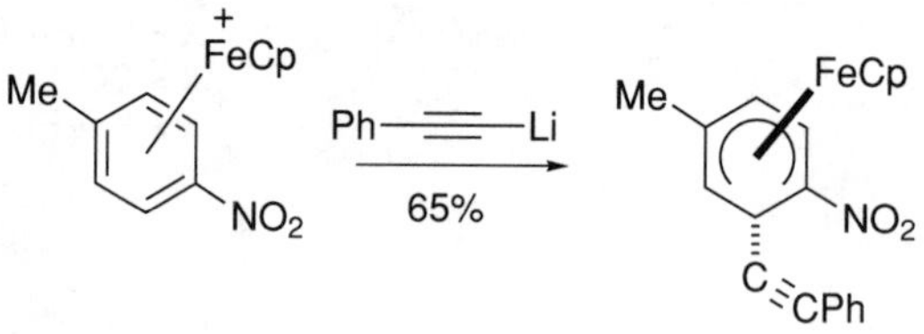

Me, $\overset{+}{FeCp}$, NO_2 —[Ph—≡—Li, 65%]→ Me, FeCp, NO_2, C≡CPh (6.20)

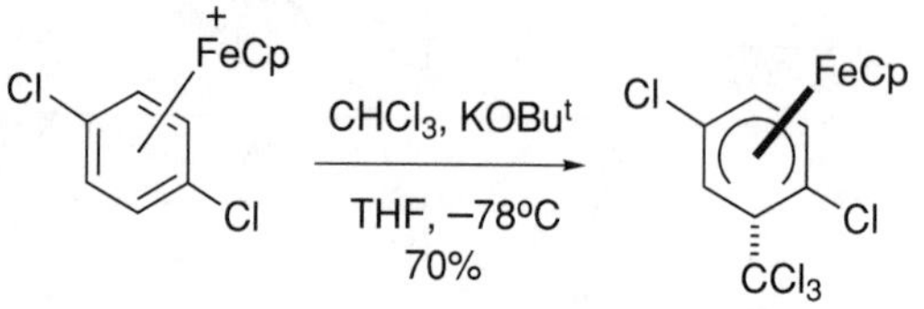

Cl, $\overset{+}{FeCp}$, Cl —[$CHCl_3$, $KOBu^t$; THF, –78°C; 70%]→ Cl, FeCp, Cl, CCl_3 (6.21)

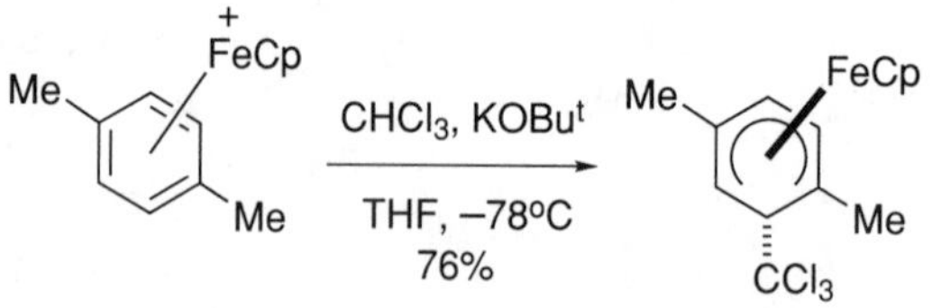

Me, $\overset{+}{FeCp}$, Me —[$CHCl_3$, $KOBu^t$; THF, –78°C; 76%]→ Me, FeCp, Me, CCl_3 (6.22)

Interestingly, the addition of trichloromethyl anion shows some reversibility, and can be driven thermodynamically[9]. At short reaction times mixtures of regioisomers are obtained during additions to complexes of *ortho*- and *meta*-xylenes, and indane. The use of longer reaction times leads to the exclusive formation of dienyl complexes resulting from additions to the sterically less hindered position (equations 6.23–6.26). A one-pot procedure has been developed in which the dienyl complexes are converted into aromatic compounds by *in situ* oxidation with iodine.

FeCp, Me, Me → Me, Me, CCl_3

1) $CHCl_3$, $KOBu^t$; THF, –78°C, 4h; 2) I_2; 55% (6.23)

+ FeCp, Me, Me → Me, CCl_3, Me

1) $CHCl_3$, $KOBu^t$; THF, –78°C, 4h; 2) I_2; 50% (6.24)

+ FeCp → CCl_3

1) $CHCl_3$, $KOBu^t$; THF, –78°C, 24h; 2) I_2; 45% (6.25)

+ FeCp → CCl_3

1) $CHCl_3$, $KOBu^t$; THF, –78°C, 1h; 2) I_2; 65% (6.26)

In the absence of reversibility during nucleophile additions to alkyl-substituted arene-FeCp complexes the formation of regioisomeric products is observed. This problem is amply illustrated in equation 6.27[10], by comparison with the result shown in equation 6.26. The importance of steric effects in the arene itself is illustrated in equation 6.28, where the presence of the *gem*-dimethyl groups leads to excellent regiocontrol.

+ FeCp → (Nucleophile) FeCp, R + FeCp, R (6.27)

$NaBH_4$	48:52	(70%)	R = H
$LiBEt_3H$	71:29	(97%)	R = H
BuLi	53:47	(60%)	R = Bu
Bu^tLi	50:50	(80%)	R = Bu^t
PhLi	64:36	(96%)	R = Ph

Nucleophile

Nucleophile	Yield	R
$LiBEt_3H$	95%	R = H
BuLi	58%	R = Bu
PhLi	59%	R = Ph

(6.28)

6.2.3 Demetallation of dienyl-FeCp complexes

This transformation has been mentioned briefly in the preceding section. In a number of cases oxidation of the dienyl-FeCp complexes that result from nucleophile additions can be accomplished by treatment with 2,3-dichloro-5,6-dicyano-1,4-benzoquinone (DDQ) in acetonitrile as solvent. This method, which appears to be quite efficient for demetallation of cyanide addition products, has been used for other adducts, and leads to the formation of the substituted arene. Some examples are given in equations 6.29–6.33[6–8]. Ceric ammonium nitrate under buffered conditions[11] has also been used to effect this demetallation.

DDQ, CH_3CN, rt, 30 min

R = Me (69%)
R = OMe (74%)
R = OPh (68%)

(6.29)

DDQ, CH_3CN, rt, 30 min

R = H, Z = CH_2NO_2 (70%)
R = H, Z = CH_2CN (73%)
R = H, Z = CCPh (77%)

(6.30)

DDQ, CH_3CN, rt, 30 min, 78%

(6.31)

DDQ, CH_3CN, rt, 30 min, 78%

(6.32)

FeCp, CCl$_3$ —[DDQ, CH_3CN; rt, 30 min; 55%]→ CCl$_3$ (6.33)

Iodine has also been used to oxidize the dienyl-FeCp intermediates from nucleophile addition, and has been used in a one-pot procedure (see earlier, equations, 6.23–6.26). These procedures are not general, however, and in a number of cases scission of the C–C bond (equation 6.30) occurs to give aromatic material in which the substituent, that was introduced during nucleophile addition, has been lost. There is still a great deal of work to be done in this regard. Also, there are no reports (of which this author is aware) of conversion of dienyl complexes into substituted cyclohexadienes or cyclohexenones. Such interconversions have been developed in the arene-$Cr(CO)_3$ and arene-$Mn(CO)_3$ area, and add an extra dimension to this chemistry. A determined effort is definitely needed in the area of dienyl-FeCp chemistry in order to identify methods for effecting this useful transformation.

6.2.4 Nucleophilic substitution reactions

One example of this reaction was shown in equation 6.17. Transition metal complexes of halobenzenes in general allow nucleophilic substitution reactions to occur. The order of reactivity is chlorobenzene-$Mn(CO)_3$ > chlorobenzene-FeCp ~ 2,4-dinitrochlorobenzene>chlorobenzene-$Cr(CO)_3$. Generally fluoroarene complexes are more reactive than the corresponding chloroarene complexes, but they are usually more difficult to prepare. Bromo- and iodoarene complexes are not very useful for effecting S_NAr reactions. In general the substitution reactions work well with less reactive nucleophiles, since these can add reversibly to the arene complex. When this condition is not met, simple addition *ortho-* to the halide occurs, as outlined in equations 6.9, 6.15, and 6.21. Malonate enolates, amines, alkoxides and phenoxides are particularly effective for nucleophilic substitutions (equations 6.34–6.35)[12]. Leaving groups other than halide can also be used in certain cases. For example, xanthone complexes undergo nucleophilic substitutions that lead to opening of the oxygen ring (equation 6.36)[2].

R, $\overset{+}{\text{FeCp}}$, Cl —[$R'CH(CO_2Et)_2$; $KOBu^t$, DMF, THF; reflux, 10h]→ R, $\overset{+}{\text{FeCp}}$, $CR'(CO_2Et)_2$ (6.34)

R = H, R′ = Et (70%)
R = *o*-Me, R′ = Et (67%)
R = *p*-Me, R′ = Me (89%)
R = *o*-Cl, R′ = Me (72%)
R = *m*-Cl, R′ = Me (79%)
R = *p*-Cl, R′ = Me (76%)

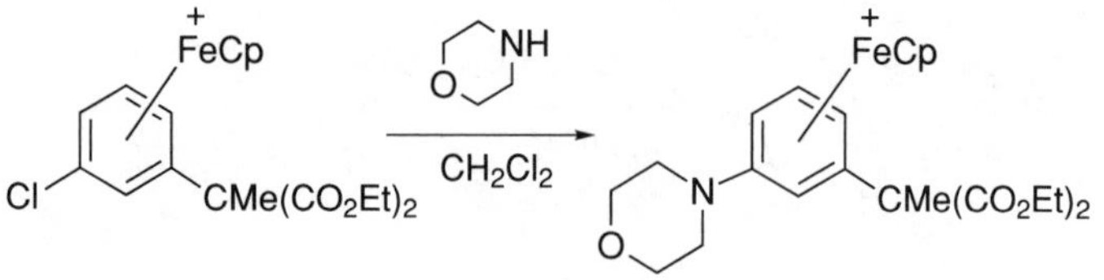

(6.35)

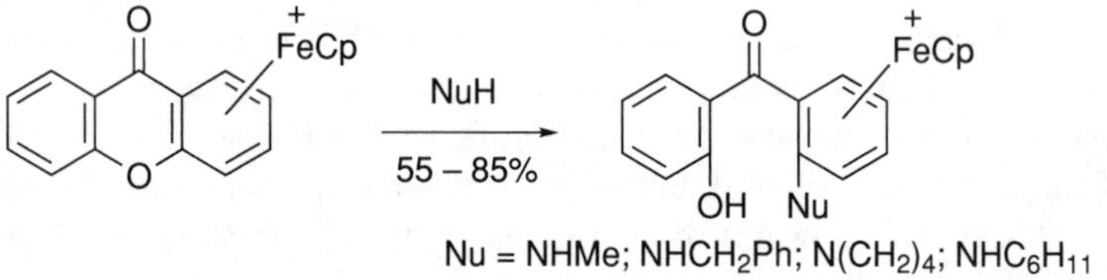

(6.36)

Nu = NHMe; $NHCH_2Ph$; $N(CH_2)_4$; NHC_6H_{11}

Excellent selectivity can be achieved during these reactions. Three examples are shown in equation 6.34, where one chloride is displaced from *ortho*-, *meta*-, or *para*-dichlorobenzene complexes by malonate. The same selectivity can be achieved using phenoxide and amine nucleophiles. It should be noted that electron-donating substituents such as ethers or amines deactivate the arene toward nucleophilic attack, thereby leading to even greater potential for selectivity during these reactions. Some illustrative examples are given in equations 6.37–6.42[13].

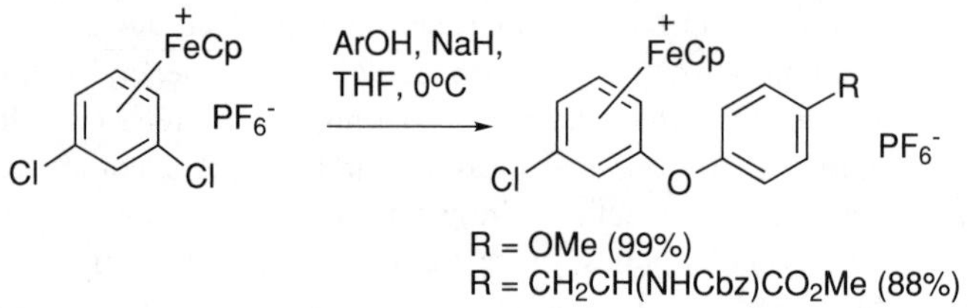

(6.37)

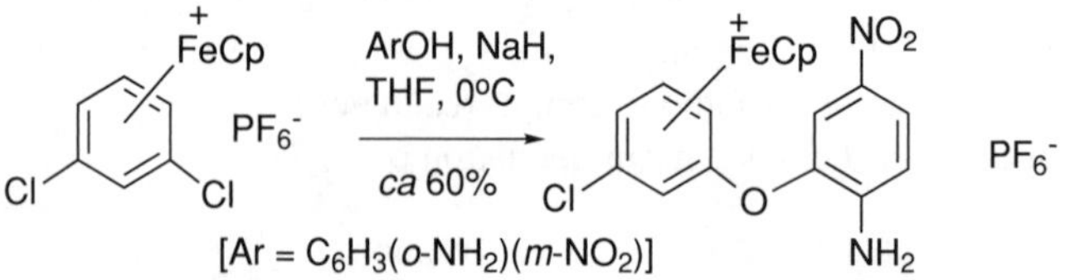

R = OMe (99%)
R = $CH_2CH(NHCbz)CO_2Me$ (88%)

(6.38)

[Ar = C_6H_3(*o*-NH_2)(*m*-NO_2)]

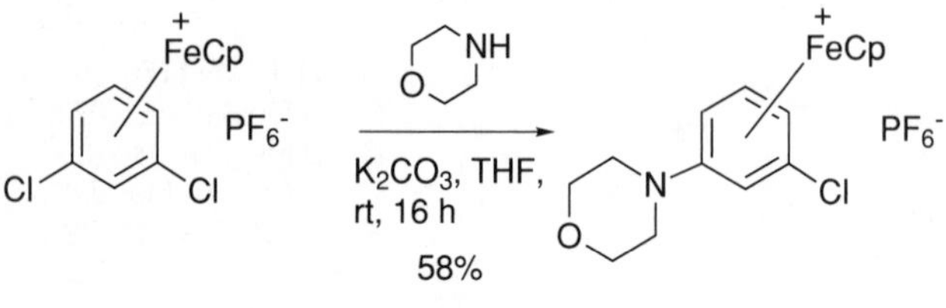

(6.39)

Prolinol, K_2CO_3, THF, rt, 99% (6.40)

(R)-(+)-PhCH(Me)NH_2, 55% (6.41)

Prolinol, K_2CO_3, THF, rt (6.42)

Double selectivity is observed for reactions of aminophenols and hydroxy amines. For example, treatment of 2-hydroxy-4-nitroaniline with sodium hydride leads to selective deprotonation of the phenolic OH to generate the substituted phenoxide which then displaces chloride from 1,3-dichlorobenzene-FeCp (equation 6.38). On the other hand, prolinol reacts on nitrogen, reflecting the greater nucleophilicity of aliphatic amines vs. alcohols (equations 6.40 and 6.42). Reactions of chiral nucleophiles with 1,2- and 1,3-dichlorobenzene complexes can give rise to diastereomeric monosubstitution products, owing to the planar chirality present in their arene-FeCp systems. With the 1,3-dichlorobenzene complex no diastereoselectivity is observed in this reaction, but 1,2-dichlorobenzene-FeCp gives approximately 1.5:1 diastereoselectivity during its reaction with (*R*)-(+)-1-phenylethylamine (equation 6.41)[14] or L-prolinol[15]. While this represents only modest selectivity, it suggests that this could be ultimately used as a new approach to asymmetric synthesis. The aryl ether complexes shown in equation 6.37 undergo displacement of the remaining chloride on reaction with a variety of other nucleophiles (equations 6.43 and 6.44)[13] although the amino-substituted derivatives (equations 6.39–6.42) are deactivated and are more difficult to substitute further. Reactions of 1,2-dichlorobenzene-FeCp with dinucleophilic substrates can also lead to displacement of both chlorides to give cyclized products (equation 6.45)[16].

Ar^1O–C_6H_4(Cl)–$FeCp^+$ PF_6^- $\xrightarrow[85\text{–}87\%]{Ar^2ONa,\ THF,\ 0^oC}$ Ar^1O–C_6H_4(OAr^2)–$FeCp^+$ PF_6^- (6.43)

Ar^1O–C_6H_4(Cl)–$FeCp^+$ PF_6^- $\xrightarrow[72\text{–}76\%]{NaCH(X)CO_2Me}$ Ar^1O–C_6H_4($CH(X)CO_2Me$)–$FeCp^+$ PF_6^- (6.44)

X = SO_2Ph or CO_2Me

$FeCp^+$ (Cl, Cl) PF_6^- + HX, HY, R $\xrightarrow{K_2CO_3}$ $FeCp^+$ (X, Y, R) (6.45)

X = Y = O, R = H
X = S, Y = O, R = H
X = Y = S, R = Me
X = NH, Y = O, R = H
X = NH, Y = S, R = H

The chemistry outlined here provides methodology for the selective formation of di- and triaryl ethers under exceptionally mild conditions compared to the standard Ullmann coupling reaction[17–20]. In order for this to be useful, methods are required for efficient decomplexations that do not lead to destruction of sensitive functionality introduced during the coupling reactions.

6.2.5 Decomplexation of arene-FeCp derivatives

In most of the earlier studies on these compounds, disengagement of the metal has been accomplished by pyrolysis, either *via* sublimation or by simply heating in a donor solvent such as DMF or DMSO. Although this is satisfactory for robust molecules, there are obvious problems when delicate functionality is present. A much more general method for decomplexation of arene-FeCp cations is by irradiation with sunlight or UV in solvents such as acetone, acetonitrile or methylene chloride[2]. This method has been used to effect decomplexations of aryl ether derivatives that have protected amino acid side chains without racemization or loss of protecting groups, (equations 6.46–6.49)[13].

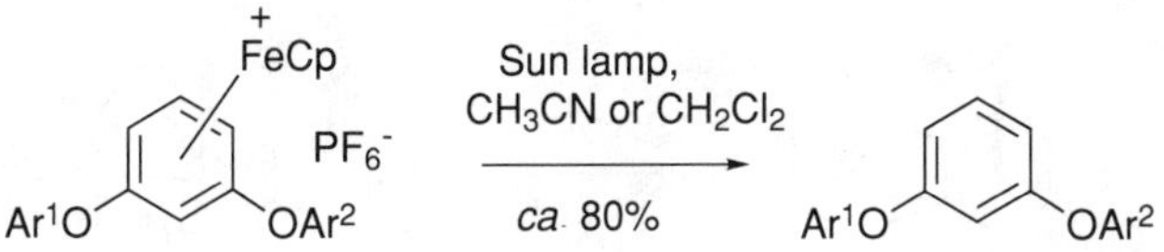

(6.46)

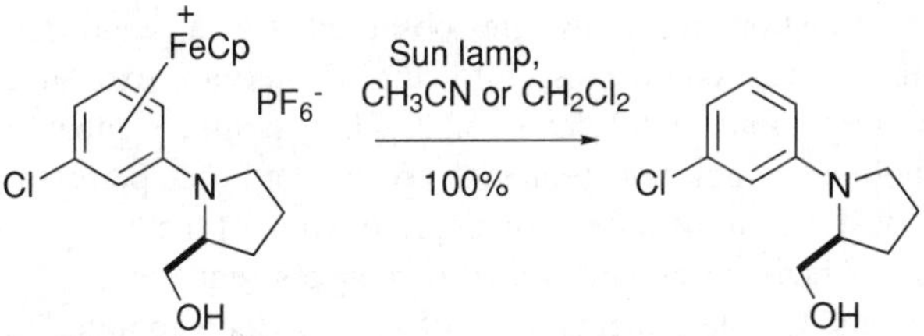

(6.47)

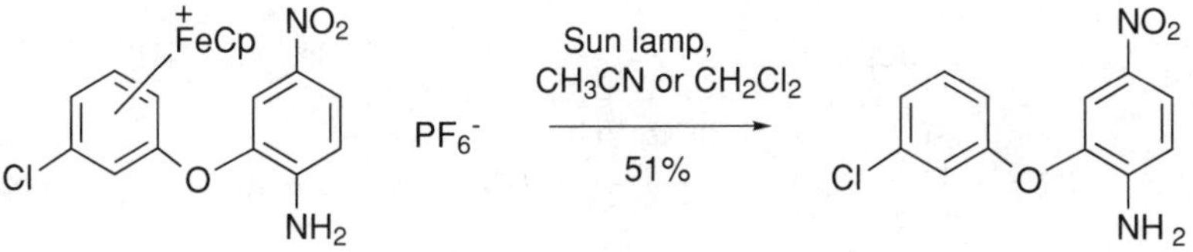

(6.48)

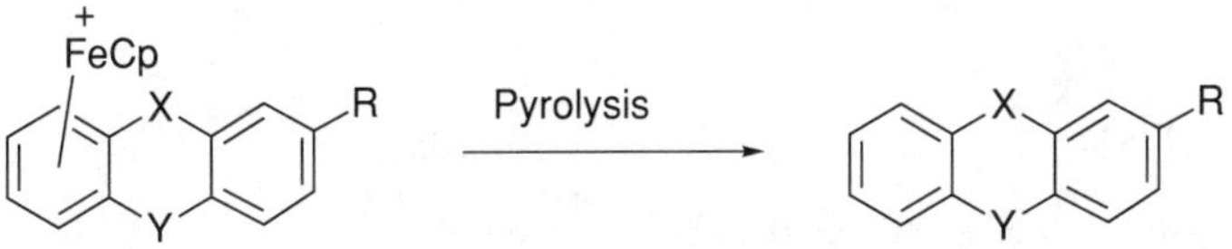

X, Y and R as in equation 6.45

(6.49)

6.2.6 Deprotonation and subsequent reactions at benzylic positions

Treatment of the FeCp complex of fluorene with potassium t-butoxide (equation 6.50) gives a blue zwitterionic complex, which is thermally stable and has been subjected to structure determination by X-ray crystallography[21,22]. This complex provided the first example of deprotonation of the benzylic positions of arene-FeCp derivatives. Resonance stabilization as shown in equation 6.50 gives some cyclohexadienyl character to the molecule, as evidenced by the interplanar angle of 11°. The anion reacts with electrophiles with complex stereoselectivity to give arene-FeCp complexes substituted at the benzylic position (equation 6.51).

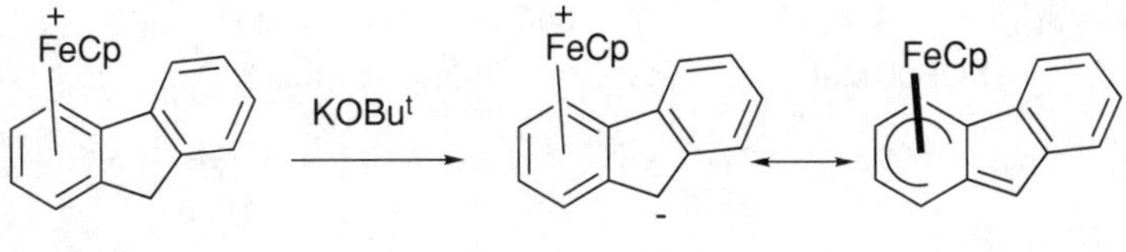

(6.50)

CH_3I (6.51)

Similar deprotonation reactions are observed for a wide range of alkyl-substituted arene-FeCp complexes, allowing deuterium exchange as well as homologation (equations 6.52 and 6.53)[2]. Tetralin (equation 6.54) and tetrahydroquinoline complexes (equation 6.55) can be permethylated at the benzylic positions[10]. In the latter case deprotonation of the NH group occurs first, a characteristic of heteroatom-substituted complexes (equations 6.58–6.60). Steric hindrance prevents complete methylation of the *o*-xylene complex (equation 6.56) but the mesitylene complex can be permethylated (equation 6.57).

NaOD, D_2O, 85°C, 15 h (6.52)

$KOBu^t$, RX (6.53)

RX = $PhCOCl$; CH_3I; Me_3SiCl; Cl_2; Br_2; I_2; CO_2; CS_2; $Mn(CO)_5Br$; $CpMo(CO)_3Br$; $CpFe(CO)_2Cl$;

50 equiv. $KOBu^t$, MeI, 64% (6.54)

$KOBu^t$, MeI, 50%; $KOBu^t$, MeI (6.55)

$\xrightarrow{\text{KOBu}^t\text{, MeI}}$ (6.56)

$\xrightarrow{\text{Excess KOBu}^t\text{, MeI, THF, rt}}$ (6.57)

$\xrightarrow[\text{2) Excess MeI, } CH_2Cl_2]{\text{1) NaNH}_2\text{, } CH_2Cl_2}$ (6.58)

$\xrightarrow[\text{2) MeI, } CH_2Cl_2\text{, } Et_2O]{\text{1) KOH, acetone, } H_2O}$ (6.59)

$\xrightarrow[\text{Base not required}]{\text{MeI, } CHCl_3}$ (6.60)

Coupled with the methods for decomplexation of the products described in the preceding section, this chemistry leads to powerful methodology for homologating alkylbenzenes.

6.3 BIS(ARENE)IRON COMPLEXES

6.3.1 Preparation

Bis(arene)iron dications can be prepared using the conventional Fischer–Haffner method[23,24], which involves treatment of the arene with $FeCl_2$ or $FeBr_2$ in the presence of $AlCl_3$ at elevated temperature (equation 6.61). The arene may be used as solvent, or if this is a solid, an inert hydrocarbon solvent such as decalin can be employed. In some cases yields are improved by using $FeCl_3$ instead of $FeCl_2$.

$$\text{C}_6\text{H}_5\text{R} \xrightarrow[\text{80–190°C, 1–20 h}]{\text{FeCl}_2,\ \text{AlCl}_3} [(\text{RC}_6\text{H}_5)_2\text{Fe}]^{2+}(\text{AlCl}_4^-)_2 \quad (6.61)$$

Alkyl-substituted arene complexes tend to be more hydrolytically stable than simple benzene complexes, and the tetrachloroaluminates are usually converted into hexafluorophosphates which are easier to handle, being soluble in acetonitrile and acetone, almost insoluble in water and alcohols, and completely insoluble in ether, hydrocarbon solvents and THF.

An alternative method for the preparation of bis(arene)iron complexes involves treatment of the arene with a 1,1′-diacylferrocene in the presence of $AlCl_3$ at elevated temperature (equation 6.62)[25–26].

$$(\text{R}'\text{COC}_5\text{H}_4)_2\text{Fe} \xrightarrow[\text{115–130°C}]{\text{C}_6\text{H}_5\text{R},\ \text{AlCl}_3} [(\text{RC}_6\text{H}_5)_2\text{Fe}]^{2+}(\text{AlCl}_4^-)_2 \quad (6.62)$$

R′ = Me or Ph

6.3.2 Nucleophile addition reactions

Owing to their dicationic nature, it is possible in principle to add two nucleophiles sequentially to bis(arene)iron complexes. These reactions have not been as well studied as those of the arene-FeCp complexes discussed above. The first nucleophile addition occurs with complete stereoselectivity to produce an (arene)(cyclohexadienyl)iron monocation complex (equation 6.63). A second nucleophile could, in principle, attack either the arene or cyclohexadienyl ligand, and reports of both modes of addition have appeared[2,27]. In practice these reactions seem to be rather capricious, and some examples of typical nucleophile additions are given in equations 6.64 and 6.65. The most recent study[27] suggests a preference for the second nucleophilic addition to occur at the cyclohexadienyl ring to give the cyclohexadiene complexes, in accordance with the Davies, Green, Mingos rules[4]. In some cases the second nucleophile addition does not proceed at all; instead demetallation occurs (equation 6.66).

$$[(\text{C}_6\text{Me}_6)_2\text{Fe}]^{2+} \xrightarrow{\text{NaBH}_4,\ 0°\text{C}} [(\text{C}_6\text{Me}_6)\text{Fe}(\text{C}_6\text{Me}_6\text{H})]^{+} \quad (6.63)$$

Me Me Fe Me Me $]^{2+}$ 2 x PhLi; THF, –80ºC; <5% yield → Me Me Fe Me Me Ph Ph (6.64)

Me Me Me Me Me Me Fe Me Me Me Me Me Me $]^{2+}$ MeLi, THF; –100ºC; 37% → Me Me Me Me Me Me Fe Me Me Me Me Me Me Me $]^{+}$

MeLi, THF; –45ºC; 24%

Me Me Me Me Me Me Fe Me Me Me Me Me Me Me Me (6.65)

Me Me Me Fe Me Me Me $]^{2+}$ R^- → Me Me Me Fe Me Me Me H R $]^{+}$

R = CN; CH_2NO_2; $CH_2CO_2Bu^t$

R^-

Me Me Me + R Me Me Me

43 – 55% (6.66)

6.4 EXPERIMENTAL PROCEDURES

Preparation of cyclopentadienylmesityleneiron hexafluorophosphate[3]

The $CpFe(CO)_2Cl$ starting material is prepared as follows[28]: $[CpFe(CO)_2]_2$ (3.5 g, commercially available or prepared as in Chapter 2, p. 32) is dissolved in a mixture of ethanol (250 ml), chloroform (50 ml) and concentrated hydrochloric acid (7.5 ml), and air is bubbled through the solution for 3 h. The solution is evaporated to dryness under vacuum and the residue is extracted with water. The aqueous solution is filtered and the filtrate is extracted with chloroform. The chloroform extract is dried ($MgSO_4$), evaporated to dryness and the product is recrystallized from 15% petroleum ether in chloroform to afford red crystals, which decompose without melting above 87°C, in about 75 % yield. Treatment of $CpFe(CO)_2Cl$ with aluminum chloride (*ca* 1.2 equivalents) in refluxing mesitylene, under nitrogen atmosphere results in the smooth evolution of two equivalents of carbon monoxide. The cooled reaction mixture is poured into stirred ice–water and ammonium hexafluorophosphate (*ca* 2 equivalents) is added. The product is filtered off, washed with water, and dried under vacuum to give a yellow solid. Further purification can be effected by dissolving the product in the minimum amount of methylene chloride, filtering to remove inorganic impurities and reprecipitating the product with ether. The yield is usually around 40%.

Preparation of chorobenzenecyclopentadienyliron hexafluorophosphate (equation 6.2, R = Cl)

Aluminum powder (0.832 g, 32 mmol) is added to a stirred solution of ferrocene (6.0 g, 32 mmol) in chlorobenzene (100 ml) at room temperature. (The arene is used as solvent in these preparations. In the case of a solid arene, the addition can be carried out at the melting point of the arene.) To this solution is added aluminum chloride (8.0 g, 60 mmol), and the mixture is magnetically stirred and boiled under reflux overnight. (For solid arenes a reaction temperature of *ca* 130°C is recommended.) The stirred reaction mixture is then cooled using an external ice–water bath while small portions of ice are added to destroy any excess aluminum chloride, to give a total addition corresponding to 30–40 ml of water. The resultant mixture is filtered and the aqueous layer is collected and washed with ether (3×100 ml) to remove unreacted ferrocene and chlorobenzene. The aqueous phase is concentrated to *ca* 15 ml on the rotary evaporator and concentrated aqueous ammonium hexafluorophosphate (5.26 g, 32 mmol) is added. The insoluble product is removed by filtration, washed with ice-cold water and dried on the pad and then under vacuum to give [chlorobenzene-FeCp]PF_6 (4.5 g, 37% yield). Reprecipitation from methylene chloride using ether, as in the above preparation, affords pure complex (3.3 g, 27% yield).

Nucleophile addition to [benzene-FeCp]PF_6: reaction with lithium aluminum hydride (equation 6.8, R = H)

The procedure is a modification of that described in the literature for [benzene-FeCp]Br_3, since most modern preparations of arene-FeCp complexes involve the use of hexafluorophosphate counteranions. The benzene complex (3.92 g, 11.4 mmol) in 1,2-dimethoxyethane (25 ml) is treated with lithium aluminum hydride (2.0 g) in small portions. After 30 min, an excess of water (100 ml) is added carefully, and the mixture is extracted with petroleum ether (100 ml). The organic layer is washed with water, dried ($CaCl_2$ or $MgSO_4$) and concentrated. The concentrate is transferred to a column of alumina and eluted with petroleum ether. The eluate is evaporated and the residue is sublimed in high vacuum onto an ice cold probe to give cyclohexadienyl-FeCp as orange–red crystals in 50% yield, m.p. 135–136°C.

Addition of the trichloromethyl anion to arene-FeCp complexes and in situ *demetallation (equations 6.23–6.26)*[6,7]

The dialkylarene-FeCp complexes shown in equations 6.23–6.26 are prepared by using either of the methods described on p. 184. The trichloromethyl anion is prepared by addition of chloroform (5.0 ml) to a stirred solution of potassium t-butoxide (896 mg, 8.0 mmol) in THF (20 ml) under nitrogen atmosphere at −78°C. To this solution is added the dialkylarene-FeCp hexafluorophosphate (2.0 mmol), and the stirred mixture is allowed to warm to room temperature. Stirring is continued for the period of time indicated in equations 6.23–6.26 (varies according to the nature of the arene ligand). At this point the reaction mixture can be quenched with water and extracted with ether in the usual way to give the trichloromethyl-substituted cyclohexadienyl-FeCp complex (see next section). For the one-pot conversion into a trichloromethylarene, a solution of iodine (40 mg) in THF (10 ml) is added to the reaction mixture and stirring is continued for a further 10–12 h. The mixture is then diluted with ether, filtered, and the solution is washed with saturated aqueous $NaHSO_3$ (4×50 ml), then water (2×50 ml), and is dried ($MgSO_4$). The ether is removed by rotary evaporation and the residual oil is microdistilled (Büchi GKR-50 apparatus) to give the pure trichloromethyl-substituted aromatic compound (yields are given in the equations).

General procedure for the demetallation of cyclohexadienyl-FeCp complexes to give substituted aromatics (equations 6.31–6.33)[6,7]

The substituted cyclohexadienyl-FeCp complex (0.5 mmol) that is obtained from nucleophile addition (see, for example, above) is stirred in acetonitrile (10 ml) while DDQ (170 mg, 0.75 mmol) is added rapidly. After stirring the mixture at room temperature for 30 min, it is filtered through a glass frit and evaporated to dryness. The residue is dissolved in the minimum amount of chloroform and is introduced onto a short column (5 cm) packed with deactivated F-20 alumina. The column is eluted with pentane and then ether and these fractions are discarded. Elution with chloroform and then methylene chloride and evaporation of the combined fractions gives the aromatic product. This procedure appears to work well for cyano- and trichloromethyl-substituted cyclohexadienyl complexes, but is not generally applicable to a wide range af alkyl-substituted derivatives.

Nucleophilic substitution on chloroarene-FeCp hexafluorophosphates I

This procedure describes the selective monosubstitution reaction of 1,3-dichlorobenzene-FeCp using phenoxide nucleophiles shown in equations 6.37–6.42. The reaction described is for equation 6.37, R = OMe[13]. 1,3-Dichlorobenzene-FeCp hexafluorophosphate is prepared as given on p. 184. To a stirred suspension of this complex (4.99 g, 12.1 mmol) in THF (15 ml) at −78°C, is added dropwise a solution of 4-methoxyphenol sodium salt (12.1 mmol, prepared from 4-methoxyphenol and NaH) in THF (15 ml). After the addition, the reaction mixture is allowed to warm to room temperature and is quenched with water (5 ml). The resulting mixture is concentrated by rotary evaporation and the residue is taken up in methylene chloride (50 ml), the organic layer is washed with water (2 × 25 ml), dried over $MgSO_4$ and evaporated to dryness to give the product as a brown oil (5.97 g, 98.5%) which is at least 95% pure by ^1H and ^{13}C NMR spectroscopy. The ^1H NMR spectrum of this complex in $CDCl_3$ provides an excellent illustration of the effect of complexation on chemical shift; the protons on the complexed aromatic ring are shifted *upfield* relative to an uncomplexed aromatic, despite the fact that the metal carries a formal positive charge: δ 7.06 (4H, s, uncomplexed aromatic H), 6.50–6.43 (2H, m, H-4 and H-5), 6.24 (1H, s, H-2), 6.16 (1H, d, *J* 6.3 Hz, H-6), 5.15 (5H, s, Cp), 3.86 (3H, s, OMe). The use of $CDCl_3$ as NMR solvent is rather unusual for cationic metal complexes, which usually require more polar solvents (CD_3CN or d_6-acetone), and the increased solubility in the present case is a result of the presence of the non-polar aryl substituent.

Nucleophilic substitution on chloroarene-FeCp hexafluorophosphates II

This procedure describes the selective monosubstitution reaction of 1,3-dichlorobenzene-FeCp hexafluorophosphate using amine nucleophiles[13]. Superimposed on this regioselectivity is the chemoselectivity expected for amine vs. alcohol nucleophiles (equation 6.40). A mixture of 1,3-dichlorobenzene-FeCp hexafluorophosphate (413 mg, 1.0 mmol) and potassium carbonate (345 mg, 2.5 mmol) is stirred in dry THF (20 ml) at room temperature while a solution of L-prolinol (101.2 mg, 1.0 mmol) in THF (3 ml) is added dropwise over a period of 1 h. Stirring is continued at room temperature overnight. The solution is filtered and rotary evaporated, the residue is dissolved in the minimum amount of methylene chloride, and the solution is added dropwise to ether (150 ml) and set aside overnight at *ca* 5°C (refrigerator suitable for solvent storage). The yellow precipitate is removed by filtration and the powder is washed with ether and dried under vacuum to give (cyclopentadienyl)[1-chloro-3-{2-(hydroxymethyl)pyrrolidinyl}- benzene]iron hexafluorophosphate (473 mg, 99% yield). The ^{1}H and ^{13}C NMR spectra of the product are complicated by the fact that an equimolar mixture of diastereomers is formed.

Decomplexation of arene-FeCp complexes

The product from the above preparation (384 mg) is dissolved in acetonitrile (20 ml) and irradiated with a 125 W sunlamp at room temperature for 4 h. The solvent is removed under vacuum, the residue is redissolved in methylene chloride (some insoluble material may persist) and the solution is filtered through silica gel to give a colorless to pale-yellow solution. Evaporation of the solvent gives the crude product which is purified by flash chromatography or preparative TLC on silica gel (50% EtOAc in hexane) to give a pale-yellow oil (170 mg, 100% yield), R_f 0.22 (30% EtOAc in hexane).

References

1. M.F. Semmelhack, in *Comprehensive Organic Synthesis* (ed. B. M. Trost and I. Fleming), Vol. 4, Chapter. 2.4, Pergamon Press, Oxford, 1991.
2. D. Astruc, *Tetrahedron* **39**, 4027 (1983).
3. T. H. Coffield, V. Sandel and R. D. Closson, *J. Am. Chem. Soc.* **79**, 5826 (1957).
4. S. G. Davies, M. L. H. Green and D. M. P. Mingos, *Tetrahedron* **34**, 20 (1978).
5. A. J. Pearson, *Metallo-Organic Chemistry*, Chapter 9, John Wiley & Sons, Chichester, 1985.
6. R. G. Sutherland, C. Zhang and A. Piorko, *J. Organomet. Chem.* **419**, 357 (1991).
7. R. G. Sutherland, A. S. Abd-El-Aziz, A. Piorko, A. S. Baranski and C. C. Lee, *Synthetic Commun.* **19**, 189 (1989).
8. R. G. Sutherland, C-H Zhang, A. Piorko and C. C. Lee, *Can. J. Chem.* **67**, 137 (1989).
9. R. G. Sutherland, C. Zhang and A. Piorko, *Tetrahedron Lett.* **31**, 6831 (1990).
10. S. L. Grundy, A. R. H. San and S. R. Stobart, *J. Chem. Soc. Perkin Trans.* **1**, 1663 (1989).
11. A. J. Pearson, P. R. Bruhn and I. C. Richards, *Isr. J. Chem.* **24**, 93 (1984).
12. A. Piorko, A. S. Abd-El-Aziz, C. C. Lee and R. G. Sutherland, *J. Chem. Soc. Perkin Trans.* **1**, 469 (1989).
13. A. J. Pearson, J. G. Park and P. Y. Zhu, *J. Org. Chem.* **57**, 3583 (1992).
14. K. Bambridge and R. M. G. Roberts, *J. Organomet. Chem.* **401**, 125 (1991).
15. A. J. Pearson and P. Y. Zhu, unpublished results, Case Western Reserve University, 1992.
16. R. G. Sutherland, A. Piorko, U. S. Gill and C. C. Lee, *J. Heterocycl. Chem.* **19**, 801 (1982).
17. F. Ullmann and P. Sponagel, *Justus Liebigs Ann. Chem.* **350**, 83 (1906).
18. G. Soula, *J. Org. Chem.* **50**, 3717 (1985).
19. M. Tomita, K. Fujitani and Y. Aoyagi, *Chem. Pharm. Bull.* **13**, 1341 (1965).
20. H. Weingarten, *J. Org. Chem.* **29**, 977, 3624 (1964).
21. J. W. Johnson and P. M. Treichel, *J. Chem. Soc. Chem. Commun.* 688 (1972).;
22. J. W. Johnson and P. M. Treichel, *J. Am. Chem. Soc.* **99**, 1427 (1977).
23. E. O. Fischer and R. Böttchner, *Chem. Ber.* **89**, 2397 (1956).
24. J. F. Helling and D. M. Braitsch, *J. Am. Chem. Soc.* **92**, 7207, 7209 (1970).

25. D. Astruc, *Tetrahedron Lett.* 3437 (1973).
26. D. Astruc and R. Dabard, *Tetrahedron* **32**, 245 (1976).
27. D. Mandon and D. Astruc, *J. Organomet. Chem.* **369**, 383 (1989).
28. T. S. Piper, F. A. Cotton and G. Wilkinson, *J. Inorg. Nucl. Chem.* **1,** 165 (1955).
29. M. L. H. Green, L. Pratt and G. Wilkinson, *J. Chem. Soc.* 989 (1960).

Index of Compounds and Methods

B

C

E

H

I

K

L

M

Q

R

S

T